THE RISE AND FALL OF THE SOUTHERN ALPS

THE RISE AND FALL OF THE SOUTHERN ALPS

The Rise and Fall of the Southern Alps

Glen Coates

with illustrations by Geoffrey Cox

CANTERBURY UNIVERSITY PRESS

First published in 2002 by
CANTERBURY UNIVERSITY PRESS
University of Canterbury
Private Bag 4800
Christchurch
NEW ZEALAND

mail@cup.canterbury.ac.nz
www.cup.canterbury.ac.nz

Reprinted 2003, 2005

ISBN 0-908812-93-0

Designed and typeset by Richard King
at Canterbury University Press

Photographs by Glen Coates except for those
with individual acknowledgment

Printed by Rainbow Print, Christchurch, New Zealand

COVER: At 3754 metres, Aoraki/Mt Cook stands higher than all the other peaks in the
Southern Alps. However, the summit was reduced in height by 10 metres when the East Face
collapsed on 14 December 1991. A large section of the mountain can be seen missing on the
right in this photograph taken a few weeks later. The tiny figure of Bryan Moore,
first to climb the new summit, ascends the ridge on the left.
Photo M. J. McSaveney, Institute of Geological & Nuclear Sciences

TITLE PAGE: From the west the Southern Alps present an imposing barrier to weather systems
coming in from the Tasman Sea. Aoraki/Mt Cook stands tallest on the skyline.
Close to the Main Divide a large snow-collecting basin feeds snow and ice down into
the Franz Josef Glacier.
Photo D. L. Homer

CONTENTS

ACKNOWLEDGEMENTS

DURING THE WRITING of this book a number of scientists kindly spent time reviewing all or some of the chapters, and discussing the scientific detail. Their constructive feedback has helped ensure the science presented is accurate and up to date.

Very special thanks go to Dr John Bradshaw (University of Canterbury), whose knowledge of New Zealand stratigraphy, and vast experience with the Torlesse terrane has been invaluable in improving the accuracy of some chapters. It was during my student days at the University of Canterbury in the 1970s that John's lectures first kindled my interest in the mysteries of the Torlesse rocks – where did they come from, and how did they get to New Zealand? In some ways, a few of the chapters are a popular account of John's many years of research.

Special thanks also to Eileen McSaveney (Institute of Geological & Nuclear Sciences, Lower Hutt) and Dr Simon Cox (IGNS, Dunedin). Eileen reviewed the manuscript, providing many suggestions and clever ideas, and Simon's input was invaluable in clarifying some of the scientific concepts and enabling me to use or adapt some important illustrations.

I am also grateful to a number of other scientists who checked the accuracy of specific sections or supplied data relevant to a number of figures, in particular Dr Tim Stern (Victoria University of Wellington), Dr Roddy Henderson (National Institute of Water and Atmosphere, Christchurch), Dr John Beavan (IGNS, Lower Hutt), Dr Robin Falconer (IGNS, Lower Hutt), Dr Lionel Carter (NIWA, Wellington), Dr Richard Norris (University of Otago), David Bates (Transit New Zealand, Christchurch) and Dr Trevor Chinn (NIWA, Dunedin).

Iaean Cranwell of Te Rūnanga o Ngāi Tahu, Christchurch, kindly provided the Kāi Tahu account of Kā Tiritiri o te Moana. For the use of photographs free of charge, I thank Mauri McSaveney (IGNS, Lower Hutt), John Bradshaw and Doug Lewis (University of Canterbury), Simon Cox (IGNS, Dunedin), and Rowan Taylor (Selwyn District Council).

Last but not least, I thank the former publishing manager at Canterbury University Press, Mike Bradstock, for his great patience during the years this book slowly came together, and CUP editor and designer Richard King for his efforts above the call of duty.

To conclude, I would like to pay a tribute to the many New Zealand geologists who, over the last century or so, have invested a good part of their lives mapping and making sense of this country's rocks. Their pioneering research has unravelled a complicated but fascinating geological history. It is hoped that this book does their endeavours justice, and gives their subject the colourful treatment it deserves.

The paintings and many of the diagrams and maps are the work of Geoffrey Cox. Computer artwork was created by Sandra Parkkali, Department of Conservation, Christchurch.

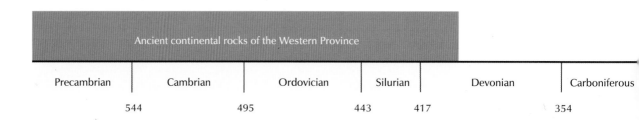

Ancient continental rocks of the Western Province					
Precambrian	Cambrian	Ordovician	Silurian	Devonian	Carboniferous
544	495	443	417	354	

THE ADVENT OF the theory of plate tectonics several decades ago revolutionised the science of geology. All over the world geologists began to use the new concept to unravel crustal movements of the past. This has been particularly true in New Zealand, where dramatic change has been going on for more than 200 million years. This book takes a look at that history, and highlights the implications of living on a plate margin – everything from earthquakes to volcanoes, and all the phenomena associated with high mountain ranges, including glaciers, rock avalanches, huge mountain storms, and raging floods.

The book is also a window into the way geologists interpret rocks and investigate the past. Unlike chemistry or physics, where researchers often have the luxury of theorems or numerical solutions, geology tends to evolve by way of observations recorded in the field, ideas and then hypotheses. Thus geologists may argue among themselves for decades over a new concept. This was indeed the case with the concept of plate tectonics last century, until it became widely accepted as a unifying theory that tied together a multitude of observations.

A trademark of geologists, then, is the 'working model' – a set of beliefs that fit the most recent observations but could be modified or replaced by an alternative as new information comes to light. Some of the illustrations in this book, too, present working models by representing simply the current view of most geologists.

For the sake of easy reading, the text has been kept free of scientific references. A glossary can be found on page 77. Papers and books that have been used in the preparation of this book are given in the references list (see page 76). The timeline below is a handy reference for the timing of events as the story unfolds. Throughout the text, both Māori and English place-names are given when a locality is first mentioned; after that, the more common name is used.

▼ **Timeline summarising New Zealand's geological history**

The main events in New Zealand's geological history are shown in this time scale, along with the intervals during which the main rock groups were formed. The colours used for rock groups match the colours used in the figures throughout this book. Refer back to this figure, chapter by chapter, for the ages of rock groups and geological periods.

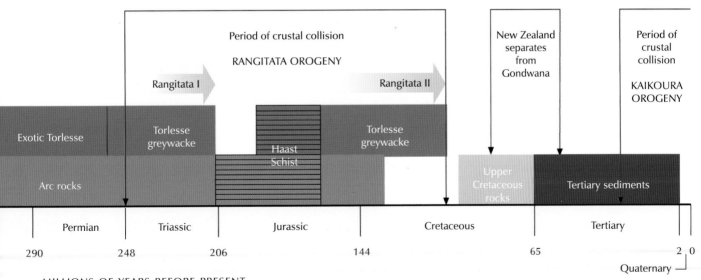

Kā Tiritiri o te Moana
The Māori Heritage

TO MĀORI, the Southern Alps are Kā Tiritiri o te Moana – 'the frothing waters of the ocean'. Ngāi Tahu, the southern Māori people, relate the creation of Kā Tiritiri o te Moana to the voyage of Aoraki and his brothers in their celestial canoe.

In the beginning there was no Te Wai Pounamu or Aotearoa. The waters of Kiwa rolled over the place now occupied by the South Island, the North Island and Stewart Island. No sign of land existed.

Before Raki (the Sky Father) wedded Papatūānuku (Earth Mother), each of them already had children by other unions. After the marriage, some of the Sky Children came down to greet their father's new wife. Among them were four sons of Raki, named Aoraki (Cloud in the Sky), Rakiroa (Long Raki), Rakirua (Raki the Second) and Rārakiroa (Long Unbroken Line). In their waka (canoe) they cruised around Papatūānuku, who lay as one body in a huge continent known as Hawaiiki.

Then, keen to explore, the voyagers set out to sea, but no matter how far they travelled they could not find land. They decided to return to their celestial home, but the karakia (incantation) that should have lifted their waka back to the heavens failed and their craft ran aground on a hidden reef.

The waka settled in the water and turned to stone and earth, forming the South Island – hence its name Te Waka o Aoraki (the Canoe of Aoraki). It listed to the east as it settled, and the higher west side became Kā Tiritiri o te Moana – the Southern Alps. Aoraki and his brothers clambered onto the high side, where they, too, were turned to stone. They are still there today. Aoraki is the mountain known to Pākehā as Mt Cook, and his brothers are the next highest peaks, Rakiroa (Mt Dampier), Rakirua (Mt Teichelmann) and Rārakiroa (Mt Tasman).

The form of the island as it now is owes much to the subsequent deeds of Tū Te Rakiwhānoa, who took on the job of shaping the land to make it fit for human habitation.

This Kāi Tahu story of the creation of Kā Tiritiri o te Moana was provided courtesy of the office of Te Rūnanga Ngāi Tahu, Christchurch.

TASMAN
SEA

ALPINE FAULT

Arthur's
Pass

SOUTHERN ALPS

Waimakariri
River

Rakaia River

CANTERBURY
PLAINS

THE SOUTH ISLAND of New Zealand has a re-markable diversity of landscapes. In the space of just several hours one can drive through fertile coastal plains, arid hinterland, rugged mountains and dense rainforest. Dominating this varied landscape, and the very reason for its diversity, are the alps, known to Māori as Kā Tiritiri o te Moana and which Europeans, remembering their own great northern mountain chain, called the Southern Alps. Stretch-ing north and south across the path of the Roaring Forties, their lofty peaks divide the South Island into two distinct geographic areas of east and west coast, each with its unique climate, landscape and life forms. Separating these two areas is the Main Divide, a line either side of which rivers drain east or west.

◀ A satellite image of the central Southern Alps. In this false-colour picture the light red-brown tones represent pasture and the dark red-brown is forest. On the West Coast the coastal plain is quite narrow and the mountains rise steeply along the Alpine Fault, a huge, surprisingly linear fracture separating two sections of the Earth's crust. To the east the mountains descend to the foothills of inland Canterbury. The cultivated patchwork of the plains extends far off the right of the picture.

Landsat 3 satellite image processed by Landcare Research, data courtesy of NASA

Despite appearances, the Southern Alps are by no means a solid, unyielding mass of rock. These moun-tains are being pushed up continually by forces deep within the Earth. Simultaneously, vast blocks of land are being shunted past one another, the occasional large earthquake reminding New Zealanders that they live on a very restless part of the Earth's crust.

At the same time as the Southern Alps are being uplifted, other natural forces are wearing the moun-tains down. Through time, rivers and glaciers have carved out a multitude of valleys and ranges. As we shall see, the walls of these valleys and the craggy, grey outcrops above reveal glimpses of a mysterious interior. Geologists who have studied these rocks over the last 150 years have unearthed a remarkable story, dating back several hundred million years ago. This book relates that story, from the first grain of sand laid down on the seafloor to the construction of one of the Earth's fastest-rising mountain systems.

▼ Looking west across Lake Tekapo from Mt Hay, the East Face of Aoraki/Mt Cook rises up behind the ranges east of the Main Divide. Sections through the Southern Alps used in Chapters 5 and 6 follow this line through Lake Tekapo and Mount Cook.

▲ In this 180-degree panorama of Mt Cook National Park, the Main Divide of the Southern Alps can be followed from Mt Sefton (far left) to Aoraki/Mt Cook, New Zealand's highest peak (3754 metres). The large lake in the centre distance has formed by the melting of ice at the terminus of the Tasman Glacier, which extends 29 kilometres upvalley. Mt Cook Airport (far right) is built on the outwash surface formed by the rivers draining the mountains.

Photograph taken from Mt Sebastopol, October 1994

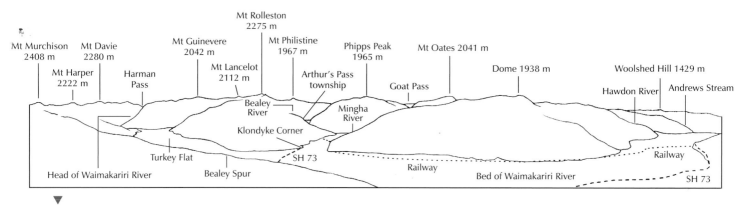

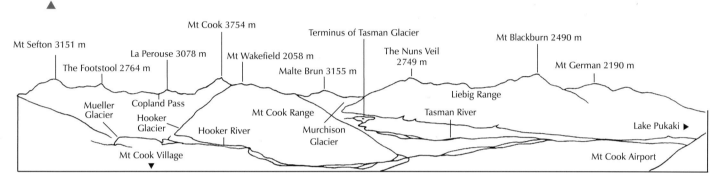

Mt Sefton 3151 m
The Footstool 2764 m
La Perouse 3078 m
Mt Cook 3754 m
Mt Wakefield 2058 m
Malte Brun 3155 m
Terminus of Tasman Glacier
The Nuns Veil 2749 m
Mt Blackburn 2490 m
Mt German 2190 m

Mueller Glacier
Copland Pass
Hooker Glacier
Hooker River
Mt Cook Range
Murchison Glacier
Liebig Range
Tasman River
Lake Pukaki ▶
Mt Cook Village
Mt Cook Airport

▼ En route to Arthur's Pass, State Highway 73 (seen far right) follows the upper Waimakariri River for many kilometres. Near Bealey the highway crosses the river and, alongside the railway line, heads up the Bealey Valley (left of centre) to Arthur's Pass township. Arthur's Pass National Park spans the horizon in this panorama.

Photograph taken from Mt Bruce, October 1995

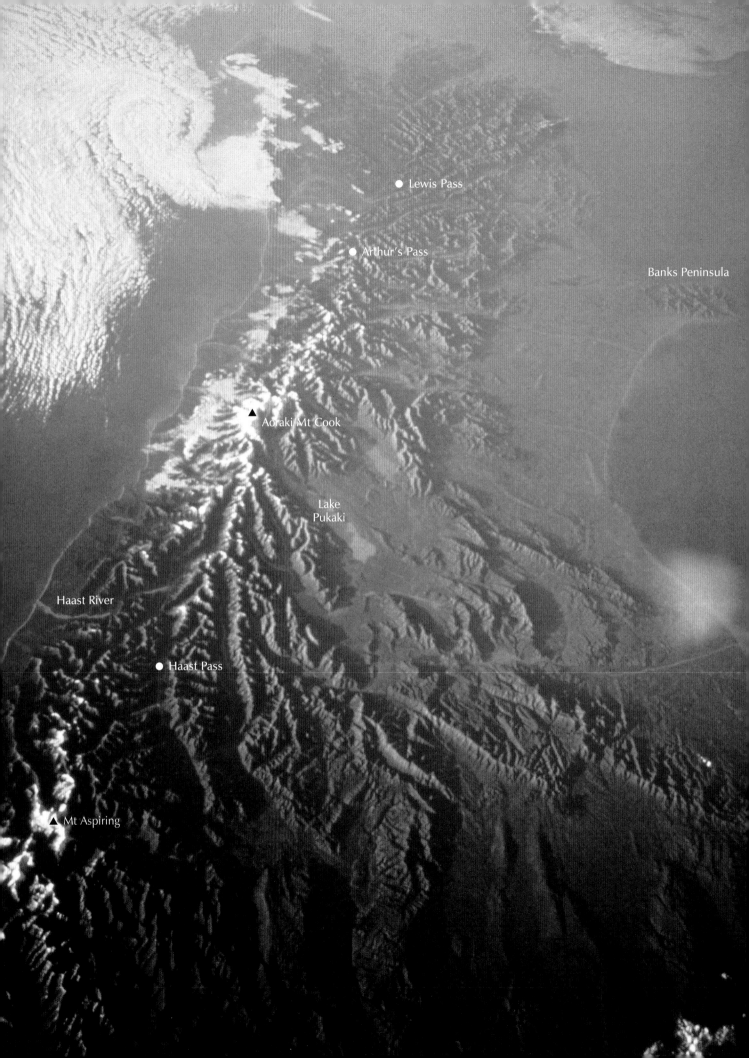

Lewis Pass

Arthur's Pass

Banks Peninsula

Aoraki/Mt Cook

Lake
Pukaki

Haast River

Haast Pass

Mt Aspiring

CHAPTER ONE

The rocks of the Southern Alps

THE SOUTHERN ALPS form the mountainous backbone of the South Island, stretching some 500 kilometres from Mt Aspiring National Park in the south to Nelson Lakes National Park in the north. The icy peaks, densely forested flanks and deep gorges of these mountains challenged many early explorers searching for routes between the east and west coasts. Māori travellers were the first to find ways through the high alpine passes in their quest for pounamu (greenstone, or jade) found on the West Coast. Later, European explorers, often led by Māori guides, sought the easiest routes to the beckoning West Coast goldfields. Even today the Southern Alps remain a considerable barrier, with highways crossing them in just three places, by way of the Haast, Arthur's and Lewis Passes.

A drive over Arthur's Pass affords an opportunity to both enjoy spectacular scenery and to look at the internal make-up of the Southern Alps. Travelling westwards towards the mountainous skyline, you gradually ascend the gentle slope of the Canterbury Plains until the highway climbs steeply into tussock grassland. The peaks of the Southern Alps, which next come into view, look grey and inhospitable, but the drab appearance of the rock belies an extraordinary sequence of events that began in the deep ocean about 300 million years ago.

Indeed, most of the Southern Alps are made up of sedimentary rocks, as can be seen in places where layers of mudstone alternate with layers of sandstone. Laid down on an ancient seafloor in horizontal layers, these beds are now nearly always inclined, folded

◀ In this NASA satellite picture the Southern Alps can be seen as a continuous chain of mountains running most of the length of the South Island.

Photo STS059-233-046, NASA Space Shuttle

Torlesse rocks are exposed below the Waimakariri Gorge bridge in inland Canterbury.
▲ The rocks consist of alternating beds of sandstone and mudstone tilted almost vertical.
▼ Small-scale laminations of sand and silt can be seen in parts of the outcrop close to the riverbed. These were formed by gentle currents flowing over the sea floor when the sediment was deposited about 200 million years ago.

or even overturned or broken into confused masses, the result of a remarkable history of mountain-building. More often, however, the original sedimentary layers are not obvious and the rock is simply hard, grey and rather uninspiringly plain. This type of rock is commonly called greywacke, a German name originally given to hard, 'dirty' grey sandstones found in the Harz Mountains of Germany. Greywacke consists of sand grains in a silt matrix, cemented into a hard grey mass.

In the eastern Southern Alps, scattered bands of black mudstone and reddish volcanic mudstone are also found, often accompanied by thin layers of chert. Further afield, small areas of limestone, marble and volcanic basalt are found. Geologists refer to this entire collection of rocks – sandstone, mudstone and these less common rock types – as the Torlesse Super-group (named after the Torlesse Range of inland Canterbury). Tens of thousands of metres thick, they were deposited for over 200 million years on an ocean floor that was once far removed from where New Zealand is today. The Torlesse rocks have, in effect, been rafted over vast distances and rammed together to create the bedrock of much of the New Zealand region. They are found from Otago to East Cape, and extend eastward beneath the sea to the Chatham Islands and south to the Auckland Islands.

The dramatic events that brought the Torlesse rocks together and deformed the original sedimentary layering have also changed, or metamorphosed, some of the Torlesse greywacke. When sedimentary rocks become buried thousands of metres deep, the enormous pressure and heat from within the Earth completely changes the original sandy and silty material, and a new type of rock, known as schist, is formed. The schist rocks formed from the Torlesse are known as the Haast Schists. They are found right across Otago and extend north as a narrow belt between the Alpine Fault and the Main Divide.

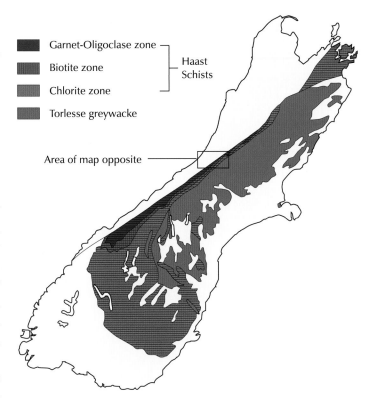

Garnet-Oligoclase zone ⎤
Biotite zone ⎥ Haast
Chlorite zone ⎦ Schists
Torlesse greywacke

Area of map opposite

▲ The rocks making up the Southern Alps consist of Torlesse rocks and schist derived by metamorphism of the Torlesse. Torlesse rocks form the ranges east of the Main Divide, and the schist rocks appear as a long narrow belt between the Divide and the Alpine Fault.

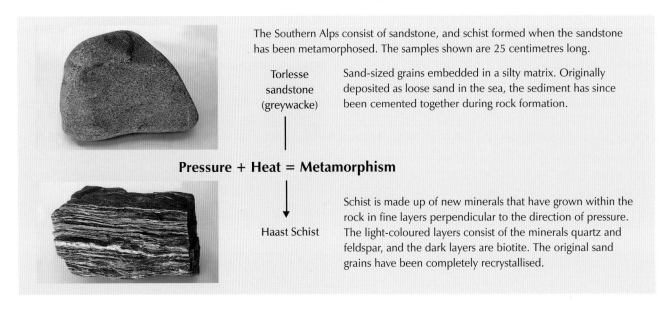

The Southern Alps consist of sandstone, and schist formed when the sandstone has been metamorphosed. The samples shown are 25 centimetres long.

Torlesse sandstone (greywacke)

Sand-sized grains embedded in a silty matrix. Originally deposited as loose sand in the sea, the sediment has since been cemented together during rock formation.

Pressure + Heat = Metamorphism

Haast Schist

Schist is made up of new minerals that have grown within the rock in fine layers perpendicular to the direction of pressure. The light-coloured layers consist of the minerals quartz and feldspar, and the dark layers are biotite. The original sand grains have been completely recrystallised.

Crossing the Southern Alps and the Alpine Fault: A highway guide

Below is a simplified geological map of the basement rocks west of Arthur's Pass. Torlesse rocks of the Southern Alps are metamorphosed to schist west of Otira, with the metamorphic grade increasing towards the Alpine Fault. West of the fault, granitic rocks form the Hohonu Range and other peaks that protrude from glacial and river sediments of the coastal plain. The metamorphic zones do not line up across the Taramakau River because of faulting along the line of the river. The Taramakau marks the line of the Hope Fault, a branch of the Alpine Fault, which continues east to Kaikoura.

If you travel east to west over the mountains along State Highway 73, and then across the Alpine Fault you can see some dramatic changes in rock type. East of Arthur's Pass, hard, grey Torlesse rocks are exposed in many of the roadside outcrops, and the same greywacke rock is seen as gravel in the mountain riverbeds and the wide riverbeds of the Canterbury Plains. However, once you reach the Taramakau River, just past Otira, the stream beds contain gravel and boulders of schist. Gravel in the Taipo River ❶ consists of both greywacke (rounder in shape) and schist (flatter), as the river's catchment area extends through both rock zones. Five kilometres on, the Big Wainihinihi River ❷ carries only high-grade schist – it drains slopes to the south-east that consist of garnet-zone schist. A few kilometers further on all the creeks such as Grahams Creek ❸ contain just granite boulders, as this is what all the high terrain west of the Alpine Fault comprises. ▶ page 18

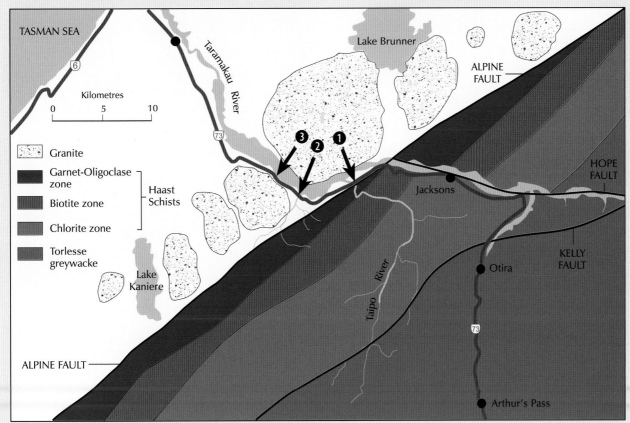

The change from Torlesse greywacke to Haast Schist can be progressively observed during the drive over Arthur's Pass (see previous page). On the Canterbury side of the Divide the road cuttings consist entirely of greywacke, and so does the gravel in the riverbeds. West of Otira, however, the mountains change to schist. As you drive towards the Alpine Fault, this schist changes progressively in character, with different minerals appearing in the rock and the metamorphic layering (schistosity) becoming more pronounced. At first, the schist has only faint layering and contains a pale green mineral, chlorite. Further west, as the rock becomes more noticeably layered, dark shiny biotite is seen throughout the rock, and finally, closer to the Alpine Fault, red garnet appears.

This Cinderella transformation from drab, grey sandstone, to low-grade schist, and then to glittering high-grade schist reflects the changes brought about by burial of the original sediments. When sediments accumulate to great thickness and are pushed deeper by forces within the crust, enormous pressure bears down upon the layers at the base of the pile. As the pressure increases with depth, so does the temperature, and the combined effect causes the mineral components within the rock to slowly change in composition. Grains of sand and mud slowly recrystallise into new minerals that grow perpendicular to the direction of pressure, forming the layering typical of slate and schist. Temperatures above 300°C and a burial depth of over 10 kilometres would have been required to change Torlesse greywacke into Haast Schist.

As the pressure/temperature conditions change with increasing depth of burial, the types of minerals that grow within rock also change. Each mineral is stable only across a limited range of pressure and temperature, and there will be a particular assemblage of stable minerals at any given depth and temperature. This is in fact what we see as we cross the Southern Alps, the different metamorphic zones having come to the surface from different depths. The low-grade, chlorite-bearing schist was formed at a relatively shallow depth, while the garnet-bearing schist to the west is the result of more intense metamorphism at greater depth.

Schist rocks come to an abrupt halt at the Alpine Fault, with completely different rocks further west. As one of the longest active faults on land in the world, this great fracture in the Earth's crust has played a key role in the uplift of the Southern Alps. However, before we look closely at the phenomenon of the Alpine Fault, let us first go back to the very beginnings of the Torlesse rocks.

▲ The Torlesse Range looms ahead dramatically while driving west across the Canterbury Plains on State Highway 73. The greywacke rocks of the Southern Alps get their name from the Torlesse Range, which was in turn named after Charles Torlesse, an early Canterbury surveyor and run-holder, and the first European to climb Mt Torlesse (1961 metres), the peak seen here on the right on the highway before Springfield.

CHAPTER TWO

Ancient sediments – the raw material of the Southern Alps

THE SEDIMENTARY ROCKS that now make up much of the Southern Alps first started to build up on the sea floor at a time when Earth was very different to the planet we know today. There were no mammals or birds, nor even any dinosaurs, and life on land was dominated by more primitive reptiles. Viewed from space, the Earth would have been unrecognisable. Vast expanses of land bore no resemblance to the continents of today, and covered completely different parts of the globe. This is because throughout the Earth's history its rigid, outer skin, which we call the crust, has been continually re-arranged by immense forces below.

In just the last few hundred million years, sections

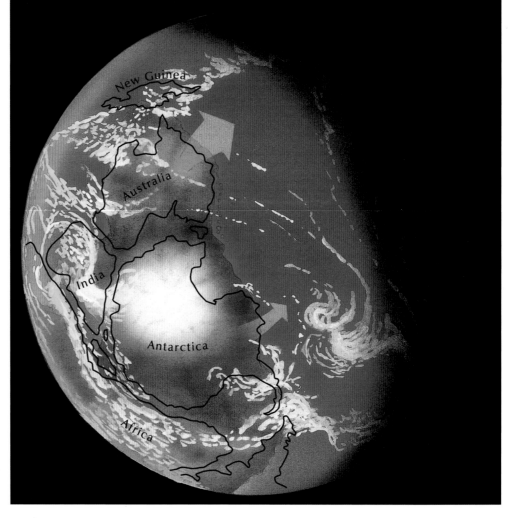

▶ This artist's reconstruction of the Earth 250 million years ago shows how our modern Southern Hemisphere continents were then united as a single supercontinent known as Gondwana. A chain of volcanic islands stretched along the east coast of Gondwana, ash and sediment eroded from these volcanoes acumulating in the adjacent sea. This sediment later became the Arc rocks of New Zealand's Eastern Province. Meanwhile, far away from the volcanic arc, sediment was being eroded from another part of Gondwana (as arrowed), sediment that later became the Torlesse rocks. The Antarctic sector of Gondwana was once thought to be the source of the Torlesse sediment, but more recent research points to north-eastern Australia.

of the crust have moved thousands of kilometres. By looking at clues on the floors of the oceans, and by examining pieces of the Earth like a jigsaw puzzle, geologists have been able to match areas that were once together and thus reconstruct the past surface of our planet.

We now know that 250 million years ago a vast continent occupied the Southern Hemisphere – the supercontinent of Gondwana. Surviving fragments of this ancient landmass are scattered across the globe, forming parts of Australia, Antarctica, India, Africa and South America. In New Zealand, Gondwana rocks form much of Fiordland, the West Coast, and west Nelson (where this country's oldest rocks and fossils are found). Some of these basement rocks date back to the Precambrian period, over 600 million years ago. Confined to the west of the South Island, they form the oldest segment of the New Zealand crust, known to geologists as the Western Province.

In contrast, the Southern Alps, the east of the South Island, and the North Island form the younger segment of New Zealand crust. Known as the Eastern Province, these rocks span a mere 200 million years of geological time, from the Carboniferous to the Cretaceous (from about 300 million to 100 million years ago). Almost all of these rocks are sedimentary.

Geologists have divided the Eastern Province into two groups, which we will call the 'Arc rocks'* and the 'Torlesse rocks'. All these rocks had their beginnings in the seas that surrounded Gondwana. Exactly where – and how – they were laid down is a mystery geologists have been unravelling for decades.

Looking down from space 250 million years ago, we would have seen a chain of volcanic islands – a volcanic arc – stretching for more than 1000 kilometres along Gondwana's eastern coastline. Some of the ash erupted from these volcanoes settled into the sea and accumulated on the sea floor. Ash and volcanic rock from on the land was also eroded and washed into the sea. This process of eruption and erosion persisted for nearly 200 million years, the sedimentary layers building up on the ocean floor. In New Zealand today, these layers are the Arc rocks of the Eastern Province. They form a large part of

---

* 'Arc rocks' is used here for simplicity as a collective term for the many formations that lie west of the Torlesse. Other authors have lumped them together as the 'Western Arc rocks' and the 'Murihiku rocks'.

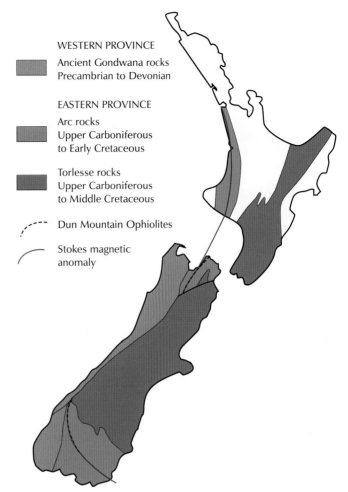

WESTERN PROVINCE
Ancient Gondwana rocks
Precambrian to Devonian

EASTERN PROVINCE
Arc rocks
Upper Carboniferous
to Early Cretaceous

Torlesse rocks
Upper Carboniferous
to Middle Cretaceous

Dun Mountain Ophiolites

Stokes magnetic anomaly

▲ New Zealand's oldest rocks are often referred to as 'basement' rocks because they form the hard foundation over which all younger sediments were laid. Basement rocks fall into two groups. Western Province rocks represent a chunk of the old Gondwana crust, and Eastern Province rocks formed from sediments that were eroded off Gondwana later. Not shown on the map, for simplicity, is the metamorphism of some of the Eastern Province rocks (the Haast Schists).

Southland, and appear again in east Nelson and South Auckland. Near Lake Wakatipu they form a small part of the Southern Alps, where they are changed into schist.

Within the Arc rocks in Southland and the Dun Mountain area of Nelson there is a narrow belt of dark greenish rocks rich in the minerals olivine and pyroxene (the Dun Mountain ophiolites). They contain a high proportion of magnesium and iron – so much iron, in fact, that they locally affect the Earth's magnetic field (the Stoke's magnetic anomaly). Geologists have traced this magnetic disturbance beyond the area where these rocks are exposed, and found that the ophiolites extend below the surface for a considerable distance to the south and north. These rocks are thought to be sections of old sea floor that

▲ Haeckel Peak (2492 metres) in the Malte Brun Range, east of Mt Cook, consists
of alternating beds of sandstone and mudstone, typical of the Torlesse rocks.
The sedimentary layers are buckled as a result of a long history of mountain uplift.

broke off and got mixed up with the volcanic and sedimentary strata.

As volcanic sediments continued to accumulate around the island arc off the coast of Gondwana, quite different sediments were accumulating in another part of the Gondwana ocean. Unlike the greenish volcanic debris from the volcanic arc, these sediments contained mainly light-coloured minerals such as quartz and feldspar, which must have been eroded from land that consisted mainly of granite. These sediments were later to become the Torlesse rocks, and like the Arc rocks their deposition continued for almost 200 million years from the end of the Carboniferous to the Early Cretaceous.

Just where on the Gondwana landmass the Torlesse material came from has been a difficult question for geologists. Because the Torlesse sediments are largely free of volcanic minerals, they must have accumulated in the ocean some distance away from the volcanic arc. The granitic rocks of what is today Marie Byrd Land in Antarctica, were once regarded as the source of the Torlesse. However, more recently the minerals in the Torlesse have shown similarities with those in the granitic rocks of Queensland, Australia, and geologists are now favouring the idea that the Torlesse came from the north (see page 20).

Regardless of where on Gondwana they were eroded from, today Torlesse rocks cover a vast area and comprise more than half of the New Zealand landmass. On land they extend from Otago to East Cape (see map on page 21), and, below the ocean, to the Chatham Islands and the Auckland Islands.

Understanding just how the huge volume of Torlesse sediments was laid down on the sea floor has been a challenge for geologists. These greywacke rocks are, in a word, monotonous – vast areas of grey, featureless sandstone, or thin beds of sandstone alternating repetitiously with thin beds of mudstone. Often the rocks are highly deformed, but there are no distinctive layers that geologists can trace over

▼ In the Otira Gorge beds of light grey sandstone and dark grey mudstone can be traced up this rock face at a steep angle. The face collapsed during the Arthur's Pass earthquake of June 1994, partially damning the Otira River and blocking the highway for nearly a week. Rock-fall protection has since been constructed over the road.

long distances to help them unravel the complicated structure. There are very few marine fossils that might tell us the age of rocks and give clues about the type of sea the sediments were deposited in – for example, whether it was shallow and inshore, or deep. Despite the huge area they cover, the Torlesse rocks have been painfully slow to yield the secrets of their distant past.

Elsewhere in the world there are other sequences of alternating sandstone and mudstone that have also puzzled geologists in the past. However, in the 1960s new clues came to light when large fans built up from layers of sandstone and mudstone were discovered on the sea floor at the foot of the continental slope. At the head of these fans, large valleys and steep-sided canyons were found cutting across the continental slope. It then became clear that sand from the continental shelf was intermittently running down these submarine canyons and onto the muddy fan surface. The world's largest deep-sea fan receives sediment from the delta of the great rivers Ganges

and Brahmaputra, and is 3000 kilometres across.

How did these fans form? Imagine the sea floor at great depth, where most of the time the water is almost still. Mud builds up slowly over a very long period, into a layer centimetres or more thick. Meanwhile, sand washed into the ocean by rivers has been building up much more rapidly at the edge of the continental shelf. Eventually this accumulation of sand becomes unstable, gives way and begins to flow down the canyon in an underwater landslide. As it gathers speed it mixes with sea water to form a dense churning mass of sediment and water called a turbidity current. Hugging the sea floor, the turbidity current moves downslope at great speed, ripping up

▼ Most of the Torlesse sediments accumulated on huge fans in the deep ocean at the foot of the continental slope. The sand and mud of the fans covered a variety of 'exotic rocks' on the old sea floor, including pillow lavas and nodules of iron and manganese. The remains of marine organisms accumulated on volcanic seamounts, and later became limestone. Sediment was also deposited on river deltas close to land.

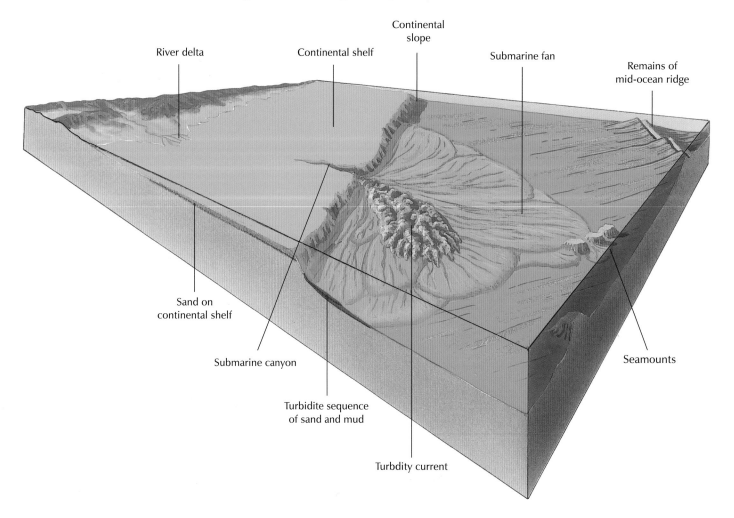

River delta

Continental shelf

Continental slope

Submarine fan

Remains of mid-ocean ridge

Sand on continental shelf

Submarine canyon

Turbidite sequence of sand and mud

Turbdity current

Seamounts

sediment from the bottom. At the mouth of the canyon the turbulent mass spreads across the surface of the submarine fan and slowly comes to rest, blanketing the muddy bottom with a layer of sand.

Geologists now recognise that the Torlesse sediments were largely formed this way. Many parts of the Torlesse are turbidites – sequences of alternating sandstone and mudstone, the mud deposited slowly in quiet, deep water, and the sand deposited rapidly from turbidity currents. Torlesse sandstone often contains lumps of mudstone ripped up from the sea floor or eroded from the canyon walls. There are also coarse sediments called conglomerate, consisting of stones embedded in mud or sand, thought to be material that came to rest within submarine canyons or in channels on a submarine fan.

That Torlesse rocks formed in the deep ocean on submarine fans also solves another puzzle – why there are so few shallow-marine fossils in most of the Torlesse strata.

While much of the Torlesse accumulated on submarine fans, there are some strata within the Torlesse

▲ From the Viaduct lookout near Arthur's Pass the imposing ridge above consists of grey sandstone so fractured and deformed that the original sedimentary layers cannot be traced. However, blocks that have fallen from the sides of the valley reveal some clues about the way that these rocks were once laid down on the sea floor. For example, this two-metre-diameter sandstone boulder photographed in the bed of the Otira River consists of black lumps of mudstone mixed up within light-coloured sandstone. ▼ The pieces of mudstone must have been ripped up from the sea floor by a powerful turbidity current, mixed with sand carried in suspension, and then deposited rapidly on the sea floor as the turbidity current came to rest. The lumps of mudstone still exhibit their internal sedimentary layering.

that must have formed by very different means. For example, at Red Rocks on the Wellington coast pillow lavas, cherts, and red and green mudstone are found within the Torlesse strata. At other localities limestone occurs, and also iron and manganese nodules. All these 'exotic' rock types are very like the materials found on parts of the ocean floor today. Nodules, for example, form by slow chemical processes on areas of the quiet sea bottom where only tiny amounts of sediment settle out. Pillow lavas are formed when molten rock erupts under the sea. Rapidly chilled by water, the lava forms a hardened outer crust. Blobs of molten material burst through the crust, forming pillow-like shapes that pile up on one another. The pillow lavas at Red Rocks could well have been erupted near an ancient mid-ocean ridge. The limestone rocks may also be related to this ocean-floor setting. For example, in today's oceans, volcanoes on the sea floor (seamounts) often build cones towards the surface. In the shallow waters above, marine organisms may flourish and their shells and calcareous body parts accumulate on top of the seamounts. This may be how the limestone in the Torlesse was formed.

Some of these unusual rock types date back into the Carboniferous, and thus date from before the Torlesse greywackes were formed. It is logical then, to view these rocks as the remains of the old ocean floor that existed some distance from Gondwana before deposition of the Torlesse sediments. Little sediment accumulated on this ocean floor until the Torlesse submarine fans spread out into the area, burying the volcanic rocks, seamounts and calcareous deposits under a great thickness of sand and mud that later became the main part of the Torlesse.

Not all the Torlesse rocks were deposited in the deep ocean, however. The greywackes of North Canterbury and southern Marlborough are the youngest Torlesse rocks, Late Jurassic to Middle Cretaceous in age. Plant fossils suggest that at least some of these strata accumulated on a river delta along the coast of Gondwana, so they seem to be the landward equivalent of the submarine-fan Torlesse.

With this picture of Torlesse deposition now in mind, we may well wonder how this diverse group of sedimentary rocks came to be part of New Zealand. We may also wonder how the Torlesse rocks came to be alongside the Arc rocks when the two came from different parts of the Gondwana ocean. To answer these questions we must look at the powerful forces that have operated within the hot interior of the planet from early in its history – forces that are still operating today.

CHAPTER THREE

Crustal collision builds new mountains

SO FAR WE HAVE SEEN that the basement rocks of New Zealand were formed at different times and in different places. The New Zealand crust is, in fact, an amalgamation of various pieces of the old continent of Gondwana and its fringing seas. All these pieces have been brought together by a phenomenon known as sea-floor spreading, a process that is constantly changing and renewing the solid outer skin of our planet. The best way to understand how sea-floor spreading has shaped New Zealand over the last few hundred million years is by looking at how the process works in today's world.

Sea-floor spreading and plate tectonics

One of the most striking features of the world's oceans is the great system of mid-ocean ridges. They mark the sites of long open wounds in the Earth's crust where magma – molten rock – has risen from the Earth's interior and erupted on the sea floor. A mid-ocean ridge is in effect a stretched-out underwater volcano, perhaps thousands of kilometres long. The upwelling of magma is caused by slow circulation within the mantle (the upper part of the Earth's

▼ Sea-floor spreading begins when magma rises from the mantle and erupts on the sea floor along a mid-ocean ridge. Convection within the mantle causes the oceanic crust to move slowly away from the ridge. Where oceanic crust collides with continental crust, the oceanic crust dives down a subduction zone. Gravity acting on the denser oceanic crust helps drive the whole process. A deep trench forms on the sea floor above the subduction zone. The melting of rock at depth causes volcanoes to erupt landward of the trench. Over time, mountains may be pushed up by the collision.

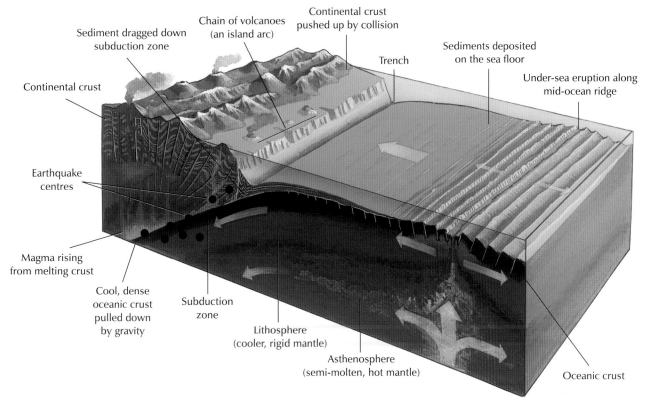

Sediment dragged down subduction zone

Chain of volcanoes (an island arc)

Continental crust pushed up by collision

Trench

Sediments deposited on the sea floor

Under-sea eruption along mid-ocean ridge

Continental crust

Earthquake centres

Magma rising from melting crust

Cool, dense oceanic crust pulled down by gravity

Subduction zone

Lithosphere (cooler, rigid mantle)

Asthenosphere (semi-molten, hot mantle)

Oceanic crust

semi-molten interior). Mantle flow, or convection, drags the crust sideways away from the ridge. With every eruption the molten material solidifies to add new basalt, a black volcanic rock, to the oceanic crust. Simultaneously the crust making up the sea floor is gradually trundled away from the rift as though on some gigantic but slow conveyor. It only moves about 10 centimetres a year, but over the great span of geological time this can translate to enormous distances. In the time since dinosaurs roamed the Earth the sea floor can move the same distance as from New Zealand to Hawaii. Thus in the vastness of geological time sea-floor spreading can change the face of the Earth.

So what happens to all this crust as it is created? Where does it go? The real action happens wherever two sections of crust are forced against one another. The denser crust literally dives beneath the more buoyant crust, and descends into the Earth's interior. This process is known as subduction, and the area where the slab of oceanic crust makes its descent is called a subduction zone. Maps of the ocean floor reveal furrows (deep-sea trenches) where subduction is taking place. These are the deepest parts of the world's oceans. As subduction proceeds, the descending crust and underlying lithosphere, which is denser than the semi-molten mantle below, is pulled downwards by gravity, and this helps drive the whole process of subduction and convection within the mantle. Eventually, the remains of the sinking oceanic crust are assimilated back into the Earth's semi-molten interior hundreds of kilometres down.

Because huge forces are involved in such a massive collision, it will not seem surprising that intense earthquake activity takes place along subduction zones. By identifying the exact position and depth of each earthquake, geologists are able to build up a three-dimensional picture of a subduction zone where it plunges into the Earth. As oceanic crust sinks down a subduction zone, heating may occur and cause parts of the rock to melt. Large bodies of molten rock may then slowly rise up through the crust above. If a path to the surface is found, volcanic eruptions will occur, resulting in either a chain of volcanic islands or a line of inland volcanoes.

The most dramatic effects of sea-floor spreading take place wherever two colliding sections of crust are carrying continental rocks. These, and the sediments that lie on parts of the sea floor, are less dense than oceanic crust and do not pass easily down a subduction zone. Instead they tend to be scraped off the underlying crust and squeezed into the collision zone. If continental materials are being carried on both colliding slabs of crust the effects may be literally earth-shattering. The rocks will be rammed into one another and gradually pushed up above sea level, eventually forming new mountains. Even the world's highest mountains – the Himalayas – were formed in this way, the Indian continent ramming into Asia about 25 million years ago.

Elsewhere on the surface of the Earth, moving sections of the crust do not collide but simply slide past one another along large faults known as transform faults. Most of these faults lie within the oceanic crust (see page 40), but some cut through continental areas, like the San Andreas Fault on America's west coast. New Zealand's own impressive example of a transform fault is the Alpine Fault.

In summary, the crust of the Earth is not a continuous solid shell but a mosaic of many rigid sections that geologists call plates (see Glossary). The boundaries between these plates are either at spreading ridges (where plates grow in size), in subduction zones (where plates are driven into the interior), or at transform faults (where plates slide past one another). The study of the movements and interactions of these plates is called plate tectonics. From studying sequences of ancient rocks, geologists now believe that sea-floor spreading has been occurring throughout the Earth's history.

Sea-floor spreading brings New Zealand together

About 250 million years ago plate tectonics began to play a major role in bringing the various parts of New Zealand together. At this time, subduction was occurring along the eastern margin of Gondwana. Magma rising from the subduction zone erupted in the sea to form the chain of island volcanoes mentioned in Chapter 2, the debris from these volcanoes later becoming the strata referred to as the Arc rocks.

Meanwhile, along another part of the Gondwana coastline, probably far to the north, Torlesse material was being eroded off the land and washed out into the deep ocean, where it came to rest on large submarine fans. As this sediment piled up, sea-floor spreading slowly moved it towards the subduction

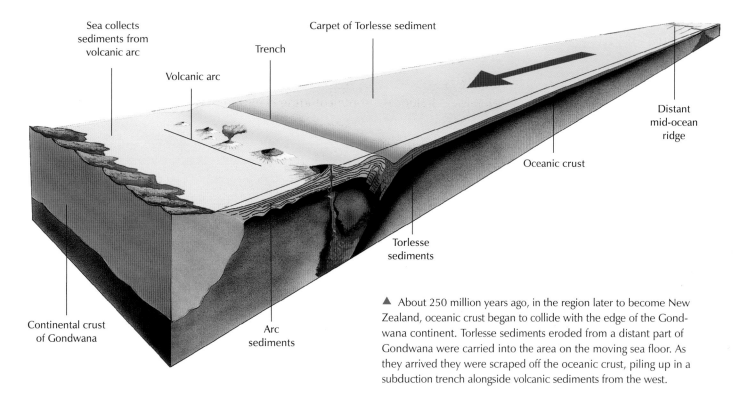

Sea collects sediments from volcanic arc

Carpet of Torlesse sediment

Trench

Volcanic arc

Distant mid-ocean ridge

Oceanic crust

Torlesse sediments

Continental crust of Gondwana

Arc sediments

▲ About 250 million years ago, in the region later to become New Zealand, oceanic crust began to collide with the edge of the Gondwana continent. Torlesse sediments eroded from a distant part of Gondwana were carried into the area on the moving sea floor. As they arrived they were scraped off the oceanic crust, piling up in a subduction trench alongside volcanic sediments from the west.

zone where the Arc sediments were accumulating. Like baggage thrown onto a slow-moving conveyor, the submarine-fan sediments were carried away. Around the beginning of the Triassic the first of this sediment arrived at the subduction trench. The heavy underlying oceanic crust descended below the edge of Gondwana, but the wet sediments were too buoyant to follow. Instead they were scraped off the crust and squeezed against the sediments in the volcanic sea to the west. This first phase of collision lasted nearly 50 million years, with more and more Torlesse material being conveyed into the collision zone.

These events occurred so long ago that the spreading ridges and sea floor that carried the Torlesse sediments no longer exist. As a result, working out the movements of the plates that caused the collision is largely guesswork. To use a modern analogy, it is as though the skid marks at the scene of a crash have disappeared, and we can no longer tell where the vehicles came from before impact.

The clues left behind in the rocks, however, suggest that by the end of the Triassic, about 200 million years ago, the sedimentary strata had formed a crumpled mass tens of kilometres thick. As more and more Torlesse sediment arrived, geologists believe there was a huge 'log jam' in the subduction process. Like baggage piling up at the end of an out-of-control

conveyor, the rocks were squeezed together, breaking and crumpling up. As the compression intensified, the strata were then slowly pushed above sea level, creating a new area of land east of the Gondwana coastline. This was just the first stage of a long period of collision and uplift that geologists refer to as the Rangitata Orogeny – named after the Rangitata River, where Torlesse rocks are well exposed. The word 'orogeny' is the geological term for a mountain-building episode.

We can only guess at just how large and mountainous this newly created land was. We do know, however, that the collision caused a great thickening of the crust, with the base of the sedimentary stack pushing down into the top of the mantle. This subjected the deeply buried rock to the high pressures and temperatures that metamorphosed them into the Haast Schists (see page 16). These rocks now form a large part of Otago and extend a long way north as a narrow strip between the Main Divide and the Alpine Fault. They also appear in the Marlborough Sounds and lower North Island. The boundary along which Torlesse and Arc rocks were brought together during collision runs through the schist in Otago and Marlborough.

Geologists are able to tell the age of schist rocks by a technique called isotopic dating, which meas-

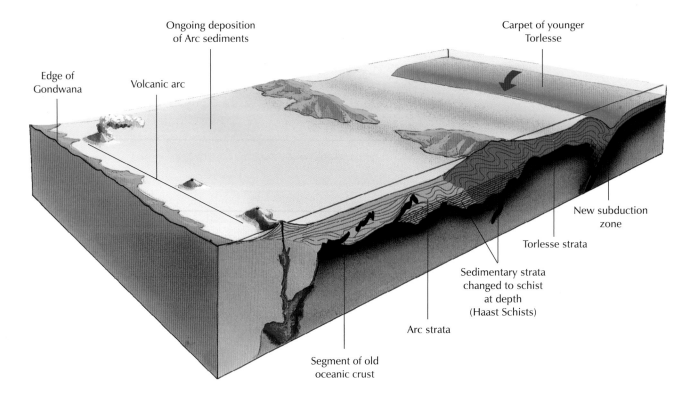

Ongoing deposition
of Arc sediments

Carpet of younger
Torlesse

Edge of
Gondwana

Volcanic arc

New subduction
zone

Torlesse strata

Sedimentary strata
changed to schist
at depth
(Haast Schists)

Arc strata

Segment of old
oceanic crust

▼ About 200 million years ago, around the end of the Triassic, the huge amount of sediment that had accumulated in the New Zealand region was pushed up above sea level to form a new area of land at the edge of Gondwana.

ures when metamorphic minerals cooled during their formation. The Haast Schists are believed to have formed from about 200 million to 160 million years ago, when it seems there was a halt in sea-floor spreading and collision. The piling-up of sediments had finished so the continental crust was in effect left to stand and cool, and the Haast Schists formed at the bottom of the pile.

What happened immediately after this first collision is unclear. With no Torlesse rocks of Early Jurassic age to be found in New Zealand, no clues have been left behind. With sea-floor spreading at a stop, new sediments may simply never have moved into the region, or perhaps subduction shifted to a new site further east away from the newly created landmass.

By the Middle Jurassic, however, sea-floor spreading had resumed, with more new Torlesse sediments arriving again on the sea-floor conveyor. This time the sedimentary pile-up continued without a break for the next 100 million years. From the Middle Jurassic to the Middle Cretaceous, newly arriving greywacke sediment was continually scraped off the sea floor

and plastered onto the margin of the new land. This added to the New Zealand crust all the Torlesse rocks found today in the east of the country from North Canterbury to East Cape. These eastern greywackes look very much like the older submarine-fan greywackes to the west. However, the mineral composition of these younger Torlesse rocks is slightly different, and there is evidence that they were formed in river deltas, suggesting they were somehow carried in from a different part of the Gondwana ocean. How this took place remains a puzzle, but the one certain thing is that, early in the Cretaceous, the scale of the sedimentary pile-up had become enormous and crustal collision was coming to a climax.

Within the Torlesse strata is an unusual zone of mixed-up rocks known as the Esk Head Mélange. Like the scar of an old wound, this zone can be traced up the middle of the country, from North Canterbury through Marlborough and on into the North Island. It is sandwiched between the older submarine-fan Torlesse and the later-arriving eastern Torlesse. It contains a great mixture of material, including blocks of old Torlesse, young Torlesse, limestone, chert and basalt, all jumbled up within squeezed mudstone (see right). Clearly something dramatic must have taken place to create this complex structure. Perhaps older Torlesse sediments were dragged down by

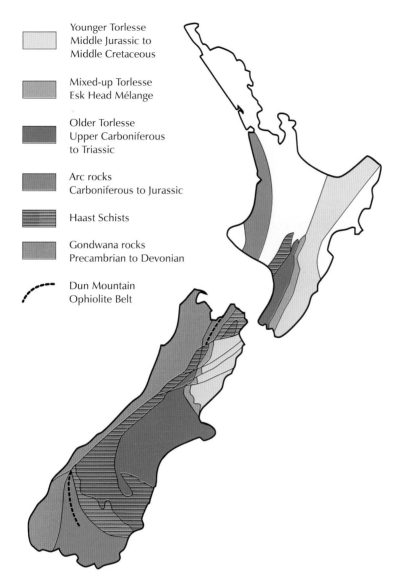

Legend:

Younger Torlesse
Middle Jurassic to
Middle Cretaceous

Mixed-up Torlesse
Esk Head Mélange

Older Torlesse
Upper Carboniferous
to Triassic

Arc rocks
Carboniferous to Jurassic

Haast Schists

Gondwana rocks
Precambrian to Devonian

Dun Mountain
Ophiolite Belt

subduction along the edge of the incoming plate, only to be squeezed back up much later and mixed with younger sediments arriving on the sea floor.

However it formed, the Esk Head Mélange reflects the intensity of collision and a pile-up of incoming sediment during the first part of the Cretaceous. Around this time parts of the thickened continental crust were again pushed above sea level. Thus began the second phase of the Rangitata Orogeny, some 100 million years after the first phase of collision and uplift that formed the Haast Schists. Geologists refer to these mountain-building events as 'Rangitata I' and 'Rangitata II'. The Haast Schists that were formed at depth during Rangitata I were pushed up by Rangitata II, raising them to a high level within the crust in the south. Thus today the Haast Schists can be seen at the surface across much of Otago. The band of schist that appears along the Alpine Fault is the result of much later crustal movement.

◀ The basement rocks of the Eastern Province can be subdivided into several 'packages' that were added one after the other to the old Gondwana crust. First to be added were the Arc and Torlesse rocks in the west, parts of which later became the Haast Schists. Joining the region later were the Esk Head Mélange and younger Torlesse in the east of the country.

◀ The Esk Head Mélange is well exposed in the face of Esk Head, inland North Canterbury. The slopes of the mountain consist of a great jumble of rock. Two large blocks of rock can be seen in this photo – a block of Triassic limestone (centre) and a block of dark pillow lavas (above and right of the limestone). They are sitting within a mixture of broken Torlesse sandstone and darker mudstone as exposed on the washed surface of a boulder ▲.

Photos J. D. Bradshaw

About 105 million years ago, after 150 million years of collision and mountain-building, sea-floor spreading finally came to an end and the Rangitata Orogeny was all over. Just how the sea-floor conveyor was 'switched off' we cannot be sure. One idea, as expressed below, is that the spreading ridge carrying the young Torlesse to New Zealand may have died out when it reached the very trench it had created.

At the end of the Rangitata Orogeny New Zealand still did not exist as a separate body of continental rock. It remained welded to the edge of Gondwana as an amalgamation of Western Province rocks and the younger Torlesse and Arc rocks of the Eastern Province. However, now that the New Zealand crust was no longer being rammed into Gondwana, a very different stage in this country's evolution was beginning.

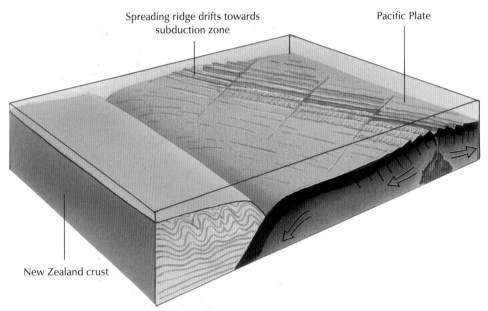

In the Middle Cretaceous the spreading ridge itself is thought to have drifted towards its own subduction trench ▲ and then simply died out ▼, bringing crustal collision to an end.

Based on ideas in Bradshaw, 1989

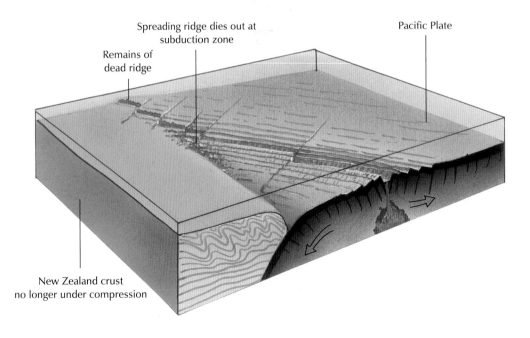

CHAPTER FOUR

A landmass declines – the sea invades

ALTHOUGH THE TORLESSE and Arc rocks are found across a large part of New Zealand, geologists now know that the same types of rock also extend across a much broader area. Deep-sea drilling, dredging and geophysical studies have shown that similar continental rocks continue below the ocean floor a great distance away from the New Zealand coastline, as far away as New Caledonia and the Chatham Islands. Around the edge of this wide area the ocean suddenly gets deeper, marking the change from continental crust to oceanic crust. This broad platform of rock is in effect a submerged continent, the drowned remains of ancestral New Zealand. The New Zealand we know today is just the highest part, poking out of the ocean.

At the end of the Rangitata Orogeny, 100 million years ago, ancestral New Zealand was a large mountainous area stretching several thousand kilometres along the eastern margin of Gondwana. What happened after this can be deciphered from the rocks that were laid down subsequently over the top of the Torlesse and basement rocks. The new strata cover the periods of time known as the Late Cretaceous and Tertiary. Although most of these younger rocks have been removed by erosion, many isolated pockets have survived (see map on next page). From these, geologists have been able to piece together the story of New Zealand's development since the Rangitata Orogeny.

New Zealand in the Cretaceous

During the first part of the Cretaceous, as Rangitata uplift came to an end, the New Zealand region was mountainous and probably larger than today. The Rangitata mountains were made of the same Torlesse rock as today's Southern Alps, so grey, barren peaks must have been typical of the ranges inland. This was

▼ A large platform of continental rock extends beneath the ocean around New Zealand. The edge of this platform is marked by a change to deeper water. The rocks forming this slab of continental crust are believed to be similar to the basement rocks found on land in New Zealand.

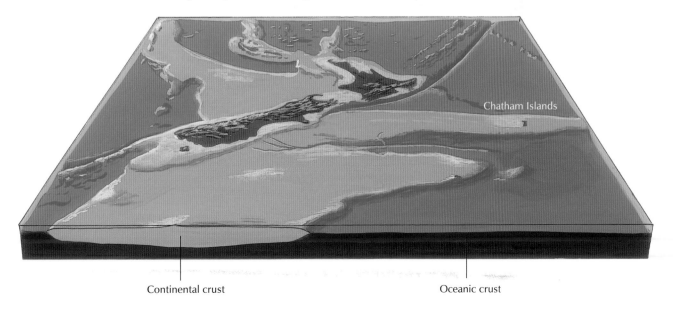

Chatham Islands

Continental crust   Oceanic crust

a period when little sediment could accumulate, as much of the area was above sea level and eroding away. The only rocks to be deposited during the Early Cretaceous were the young Torlesse strata found today along the east coast of the country – sediment that must have reached New Zealand in the dying phases of sea-floor spreading.

With crustal collision and mountain-building at an end, erosion took its toll on the Rangitata mountains. Rivers cut new valleys, washing sediment into the sea. By the Middle Cretaceous the mountains had been worn down to form extensive lowlands. Rivers flowed across fertile plains covered in vegetation. This we know from the next layers to form, which were the Cretaceous coal beds of the South Island, some of which are mined today on a large scale. Coal forms from buried, compacted plant material, and the thick coal seams found in parts of Otago, Southland, Westland and Nelson tell us that vegetation flourished in fertile swamps during the Cretaceous.

◀ Pockets of Late Cretaceous rocks (99–65 miilion years old) and Tertiary rocks (65–1.8 million years old) are scattered across New Zealand's older basement rocks.
Redrawn from Riddolls, 1987

▼ By the Middle Cretaceous much of the Rangitata mountains had been eroded away. Vegetated swamps occupied lowland and coastal areas. Sediments eroded from the landmass accumulated on the sea floor to the north-east of the region.

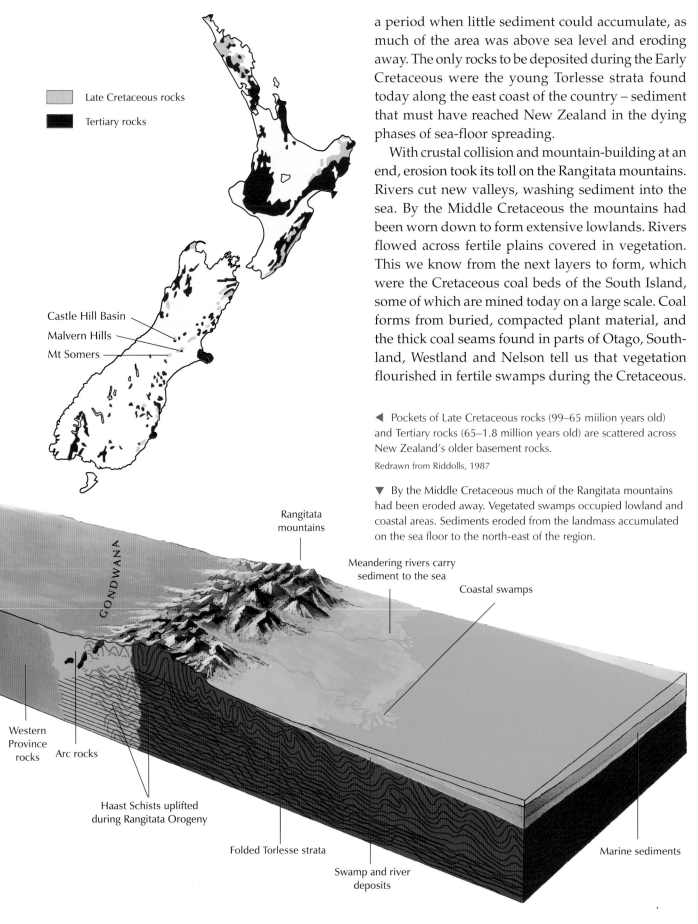

Late Cretaceous rocks

Tertiary rocks

Castle Hill Basin
Malvern Hills
Mt Somers

Rangitata mountains

Meandering rivers carry sediment to the sea

Coastal swamps

GONDWANA

Western Province rocks

Arc rocks

Haast Schists uplifted during Rangitata Orogeny

Folded Torlesse strata

Swamp and river deposits

Marine sediments

▲ These leaf fossils were once part of a plant community that grew at the edge of the Rangitata mountains in the Middle Cretaceous, about 100 million years ago. The sandstone sample is from the Clarence River, Marlborough.

I. Daniel, University of Canterbury

Coal becomes heavily compacted during its formation, so it is often difficult to recognise the original plant material from which it formed. However, layers of sandstone and mudstone sandwiched between coal seams often contain leaves and woody material. Impressions of individual leaves, along with petrified wood and microscopic fossils like pollen and spores, show that a wide range of plants grew in the Cretaceous forests of New Zealand. There were conifers, cycads, ferns and flowering plants, which made their first appearance on Earth around this time.

During the Cretaceous, animal life on Earth was still dominated by the dinosaurs, and their fossils have been found in sediments that accumulated at this time along the New Zealand coast. The bones are only fragments, but they nevertheless confirm that a variety of dinosaurs as well as pterosaurs (flying reptiles) were present.

While some of the sediment eroded from the Rangitata mountains was deposited on land, most of it accumulated in the ocean off the eastern shores of the new landmass. Thus today, marine sediments of Middle to Late Cretaceous age can be found, especially along the east side of the country. These rocks often consist of hard grey mudstones and sandstones that are almost indistinguishable from Torlesse rocks. This is because the Cretaceous sedi-

ments consist mainly of Torlesse material eroded off the Rangitata mountains. Fortunately, unlike most Torlesse rocks, Cretaceous marine sediments often contain many fossils, which tell us not only the age of the rocks but also a great deal about the sea life of the time. Bivalve molluscs had become abundant, and some of these were enormous – the shells of *Inoceramus*, for example, could reach a length of two metres. The seas teemed with fish, squid-like animals called belemnites, and ammonites – free-swimming molluscs with spiral shells up to one and a half metres in diameter. There were also many carnivorous marine reptiles such as plesiosaurs and mosasaurs, and their remains have been found in eastern Otago, North Canterbury, Marlborough and Hawke's Bay.

▲ Artist's impression of a plesiosaur (top) and mosasaur (bottom) that inhabited New Zealand seas during the Cretaceous.

Farewell to Gondwana – New Zealand goes it alone

While erosion was reshaping the Rangitata landmass during the Cretaceous, deep within the mantle the pattern of convection had shifted. Now, mantle material was starting to well upward and outward beneath the New Zealand section of Gondwana. By about 100 million years ago the whole region was starting to be pulled apart in an east–west direction. Cracks developed within the crust, along which magma could find a route to the surface. The first sign of the break-up was the eruption of new volcanoes. Today the remains of these can be seen in Canterbury and Marlborough. For example, in the Awatere and Clarence Valleys basalt is found between layers of coal, showing that eruption took place on the forested coastal areas of Cretaceous New Zealand. Further south there are rhyolite and andesite lava flows between Mt Peel and the Malvern Hills, and in the Mt Somers area.

At the same time a major split began to form in the crust west of the Rangitata mountains – the beginning of a new spreading ridge. As magma welled up within the rift, forming new oceanic crust, ancestral New Zealand began to separate slowly from Gondwana. In time, the sea flooded into the space between the dividing lands and the Tasman Sea was born. Sea-floor spreading was fully under way by 85 million years ago. New Zealand continued its journey outward into the Pacific until movement ceased about

▲ East of Cromwell in Central Otago, the Poolburn–Rough Ridge area is a good example of the Cretaceous peneplain. All the softer Tertiary rocks have been eroded away, exposing the old flat fossil landscape carved into hard schist during the Cretaceous.
Photo D. L. Homer

▼ During the Cretaceous a spreading ridge formed through Gondwana west of the Rangitata mountains. As the New Zealand continental crust separated from Gondwana and drifted away to the east, the Tasman Sea began to form. Sea-floor spreading continued to widen the Tasman Sea for 25 million years. When movement finally ceased about 55 million years ago, the New Zealand crustal block was left isolated in the south-west Pacific.

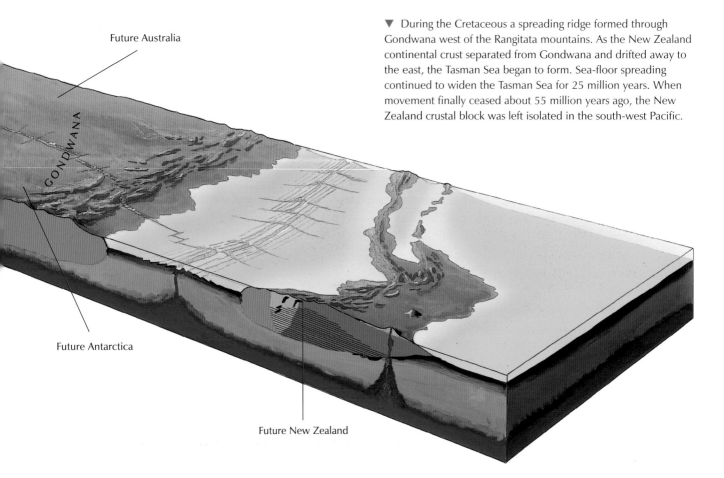

Future Australia

GONDWANA

Future Antarctica

Future New Zealand

55 million years ago, bringing to an end the opening of the Tasman Sea. For the first time the New Zealand region had an identity of its own.

On her own New Zealand begins to founder

The New Zealand landmass continued to erode down throughout the Cretaceous, even as the Tasman Sea widened and New Zealand drifted away from Gondwana. By the beginning of the Tertiary period, 65 million years ago, the land had been largely reduced to low plains. In some parts of the country today, such as Central Otago and North-west Nelson, we can still see the surface of the old flat Cretaceous landscape. Softer rocks were laid down on top during the Tertiary, but they were later stripped away to reveal the extensive flat area of hard basement rock. This exhumed fossil landscape is often referred to as the 'Cretaceous peneplain'.

Having split from Gondwana, New Zealand was now a long way from any subduction zone or spreading ridge. As a result the crust slowly began to cool and contract, becoming denser and sinking into the lithosphere. Thus, during the first 30 million years of the Tertiary the now low-lying land slowly sank

into the ocean. By the middle of the Tertiary there was very little land left above the sea. During this long period of inundation a great variety of new sediments were laid down over the old landmass. The new beds of sandstone, mudstone and limestone gradually formed a blanket across much of the region. Today, parts of the North Island are still covered with Tertiary rocks, but in the South Island most of the sedimentary blanket has been eroded away. However, around the periphery of the Southern Alps a number of patches of Tertiary sediments have been protected from erosion as a result of uplift

▲ About 35 million years ago, in the Oligocene, the New Zealand region consisted of numerous small islands – the drowned remnants of the Rangitata mountains. From this time on, the various parts of New Zealand (the present-day coastline has been added to the diagram for reference) were to be shuffled about and pushed up above sea level.

▲ The limestone beds in the middle distance are part of a sequence of Tertiary rocks in Castle Hill Basin, a triangular depression about eight kilometres wide near Arthur's Pass. The Tertiary rocks have been protected from erosion by uplift of the surrounding mountains along faults. In the distance are the greywacke rocks of the Torlesse Range.

of the surrounding terrain. Castle Hill Basin, near Arthur's Pass, is a good example, with beds of sediments deposited in the Tertiary seas preserved in a basin bounded by faults.

By studying the fossils and the types of sediment within these patches of Tertiary rocks, geologists have been able to form a picture of the seas surrounding New Zealand in the Tertiary. They have plotted the positions of ancient shorelines, which show that the sea invaded the land during the Late Cretaceous and Early Tertiary, so that by the Oligocene, about 35 million years ago, New Zealand was little more than a chain of small islands (see page 37).

Then, after the Oligocene, the shoreline around New Zealand began to pull back from the land, near-shore sediments being deposited further and further out as new land emerged. By the Pliocene, about five million years ago, large areas of land had risen from the sea. An event of major proportions was beginning to affect the New Zealand crust towards the end of the Tertiary – a new era of uplift was dawning.

CHAPTER FIVE

A new era of upheaval
– the Southern Alps are born

THE EVENTS OF THE LAST 25 million years have rotated and squeezed large areas of continental and oceanic crust, and shifted them many hundreds of kilometres. Some sections of crust have disappeared altogether as a result of subduction. It has been the quest of geologists to work out just where the various pieces of New Zealand came from and how they were shifted around.

The most important clue to this vast jigsaw puzzle lies in the rocks on either side of the Alpine Fault. Rocks of Permian age near the fault in Otago are very similar to those in the Nelson area. By splitting the South Island along the Alpine Fault and moving the two parts relative to one another, the Permian strata and other basement rocks can be matched up. This shows the configuration of New Zealand in the mid-Tertiary. Most geologists agree that the two parts of the South Island have been offset all that distance in the last 25 million years. That's a total of 480 kilometres of movement along the Alpine Fault.

The big crunch

How did a fracture as large as the Alpine Fault form in the Earth's crust, and what has driven the movement along it? To answer these questions we must look at the floor of the modern ocean. Far to the east of New Zealand lies a mid-ocean ridge that extends for thousands of kilometres along the floor of the South Pacific Ocean (see page 40). New oceanic crust being generated at this spreading ridge moves constantly away on either side. East of the ridge the crust is travelling towards the coast of South America, where it dives down into a subduction zone. On the other side of the ridge the section of crust known as

Torlesse rocks
Carboniferous to Cretaceous

Haast Schists

Arc rocks
Carboniferous to Cretaceous

Gondwana rocks
Precambrian to Devonian

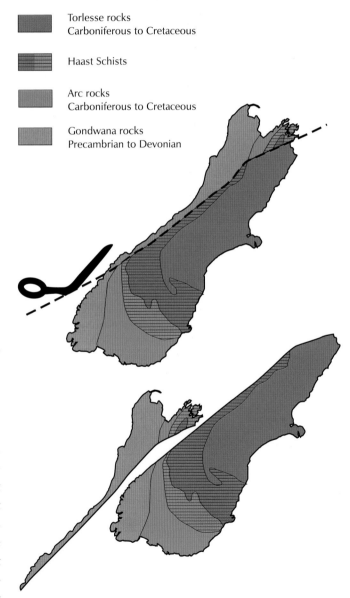

▲ If we cut the South Island in two along the Alpine Fault, we can then reassemble the parts as they used to be. By matching up the main groups of rocks in Otago and Nelson, we can see there has been nearly 500 kilometres of movement along the fault.

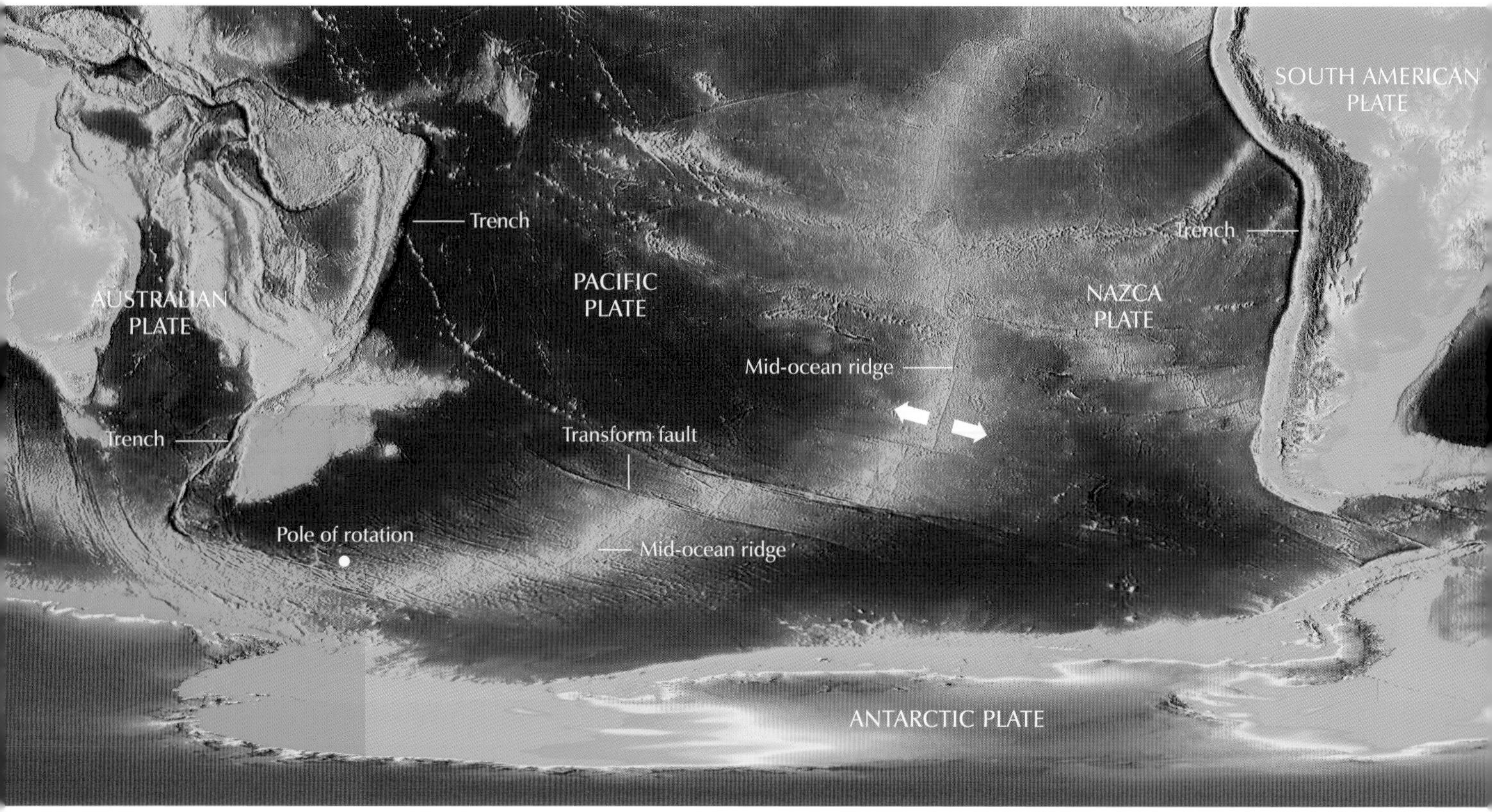
SOUTH AMERICAN PLATE
Trench
Trench
AUSTRALIAN PLATE
PACIFIC PLATE
NAZCA PLATE
Mid-ocean ridge
Trench
Transform fault
Pole of rotation
Mid-ocean ridge
ANTARCTIC PLATE

▲ The Earth's crust is divided into a number of separate rigid sections called plates (see Glossary). The boundaries between them can be seen in this bathymetric map of the South Pacific. Running thousands of kilometres up the middle of the ocean floor is a mid-ocean ridge, an irregular line along which new oceanic crust is being generated. The ridge is cut by numerous transform faults, the result of differential rates of spreading. West of the ridge oceanic crust is pushing into the South American Plate along the line of the Chilean deep-sea trench. To the east oceanic crust of the Pacific Plate is pushing into the Australian Plate – the New Zealand crust lies right across the boundary. North and south of New Zealand trenches can be seen, but, in between, the collision and movement is very complex.

Bathymetry courtesy of P. Sloss and W. H. F. Smith at NGDC of NOAA

the Pacific Plate is moving west towards the Australian Plate. New Zealand is located right in the zone where these two plate collide. A map of the relief on the ocean floor reveals the line along which collision is taking place north of New Zealand (see opposite). A deep furrow in the ocean floor, known as the Kermadec Trench, is where oceanic crust of the Pacific Plate is subducted beneath oceanic crust of the Australian Plate (see page 42). This subduction zone feeds a line of volcanoes to the west of the trench.

Near East Cape the Kermadec Trench merges into another subduction trench, the Hikurangi Trough, which runs east of Hawke's Bay and Wairarapa. Here, oceanic crust of the Pacific Plate meets continental crust, and is subducted beneath the North Island. Molten rock rising from this subduction zone feeds volcanoes in the central North Island along a line through Tongariro, Ngauruhoe, Ruapehu, Rotorua, Tarawera, and White Island in the Bay of Plenty. Lake Taupo is also part of this volcanic zone. The lake fills a crater that was the site of the most explosive volcano the world has seen in the last 5000 years.

The Hikurangi Trough peters out just north of Kaikoura. From here south, the collision becomes more complicated because both plates are made of continental crust. It is not possible for one area of continental crust to slide under another, so instead the two sections of crust are in a sideswiping collision. The side sustaining the most damage is the Pacific Plate, which is squeezed and crumpled and partly pushed up to form mountains. The entire landmass has cracked along numerous faults. The largest of these – the Alpine Fault – lies between the Pacific and Australian Plates. It is often referred to as the Alpine Fault System because it splits into a number of separate faults at the top of the South Island. The Pacific and Australian Plates are sliding past one another along this fault system, with the crust on the west side moving north relative to the east side. Sideways movement on the Alpine Fault is evident from the air in many places along its length. Topographic

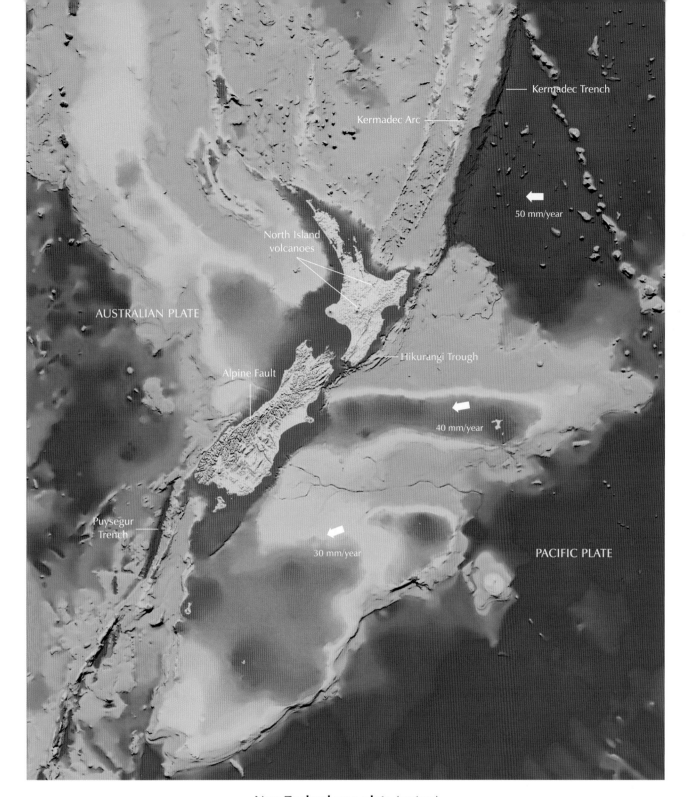

Kermadec Trench

Kermadec Arc

50 mm/year

North Island
volcanoes

AUSTRALIAN PLATE

Hikurangi Trough

Alpine Fault

40 mm/year

Puysegur
Trench

30 mm/year

PACIFIC PLATE

New Zealand on a plate (or two)

This topographic image of the ocean floor shows the line where the Pacific and Australian Plates meet. The Kermadec and Puysegur Trenches can be clearly seen north and south of the country, and the Hikurangi Trough can be seen east of the North Island running down to the Kaikoura coast. These furrows in the ocean floor mark the top of subduction zones. In this diagram sea floor coloured orange–green (shallower than 2,000 metres) approximates to areas of continental crust. Blue–purple shades (deeper than 2,000 metres) represent areas of oceanic crust. North of New Zealand, both plates are made of oceanic crust,

and collision has simply resulted in a line of volcanoes west of the subduction zone. However, east of the North Island, oceanic crust is colliding with continental crust, and magma rising from the subduction zone is more viscous, giving rise to explosive volcanoes like Mt Tarawera. Under the South Island, collision is different again because continental crust is colliding with other continental crust. Instead of subduction the crust has split along numerous faults, and mountains are being pushed up.

Bathymetry courtesy of NIWA

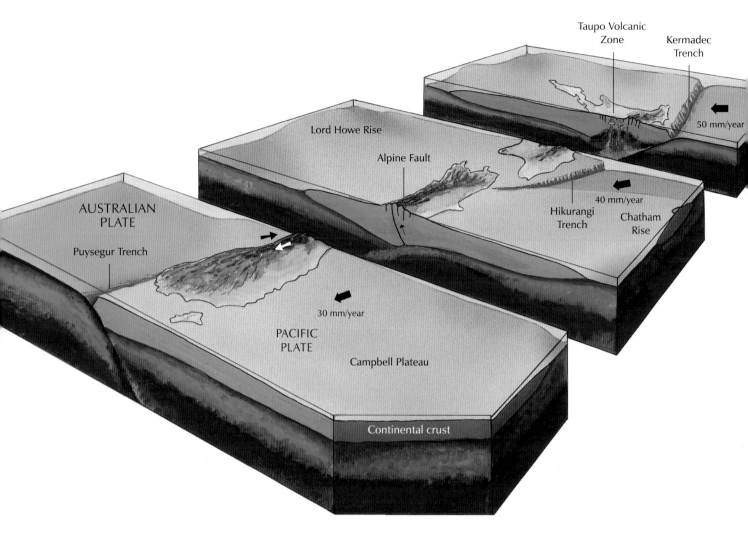

Taupo Volcanic
Zone
Kermadec
Trench

Lord Howe Rise

Alpine Fault

AUSTRALIAN
PLATE

Puysegur Trench

50 mm/year

40 mm/year

Hikurangi
Trench

Chatham
Rise

30 mm/year

PACIFIC
PLATE

Campbell Plateau

Continental crust

▲ This cut-away view of the crust shows what happens where the plates are in collision.
At the north end of New Zealand, the Pacific Plate dives down a subduction zone
beneath the North Island. South of the country subduction dips the other way,
the Australian Plate diving beneath the Pacific Plate. In the South island
the plates move past one another, mainly along the Alpine Fault.

◀ The Alpine Fault has offset the course of these
creeks seen from the air near Haupiri, east of
Greymouth. Faults, like the Alpine Fault, that feature
mainly horizontal movement, are known as
transcurrent faults. If the land on the opposite side is
displaced left to right, movement is termed right-lateral
(as opposed to left-lateral). Like most transcurrent
faults in New Zealand, movement on the Alpine Fault
is right-lateral.

Photo S. Cox

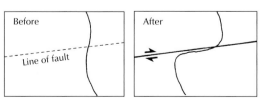

Before

Line of fault

After

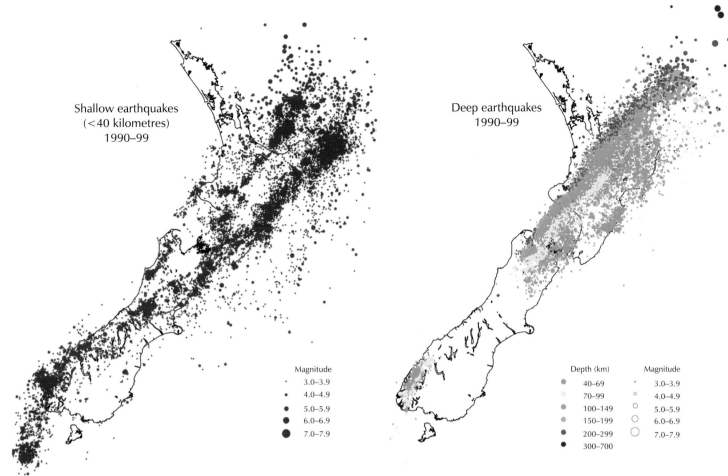

Shallow earthquakes
(<40 kilometres)
1990–99

Magnitude
· 3.0–3.9
· 4.0–4.9
● 5.0–5.9
● 6.0–6.9
● 7.0–7.9

Deep earthquakes
1990–99

Depth (km)
● 40–69
70–99
● 100–149
● 150–199
● 200–299
● 300–700

Magnitude
· 3.0–3.9
○ 4.0–4.9
○ 5.0–5.9
○ 6.0–6.9
○ 7.0–7.9

features such as streams, terraces and ridges can be seen cut through along the fault line, the two sides displaced great distances (see photo opposite).

The Alpine Fault exits the coast in Fiordland, and southward from here movement between the plates is transferred back to a subduction zone, marked on the ocean floor by the Puysegur Trench. Here, subduction faces in the opposite direction, with oceanic crust of the Australian Plate descending beneath continental crust of the Pacific Plate.

The geometry of the subduction zones at the north and south ends of the country is clearly illustrated by the pattern of earthquakes recorded beneath New Zealand. Deep earthquakes under the North Island, below a depth of 40 kilometres, become progressively deeper from east to west, reflecting the westward descent of the subduction zone. In contrast, earthquake centres under Fiordland deepen in the opposite direction, reflecting the west to east descent of the subduction zone. In the South Island, where movement between the plates is occurring along the Alpine Fault, there are hardly any deep earthquakes, which has puzzled geologists for many years. How-

▲ Earthquakes in New Zealand (1990–99) show a remarkable pattern when plotted according to their location and depth. Earthquakes that originate from depths below 40 kilometres occur under the North Island and the south of the South Island. They become progressively deeper with increasing distance from the subduction trenches. This pattern gives a three-dimensional impression of the subduction zones that descend westward under the North Island and eastward under the South Island. The lack of deep earthquakes along the Alpine Fault has raised questions among geologists and may suggest a different mechanism of movement between the plates. Shallow earthquakes are more scattered along the plate boundary and reflect the widespread release of strain within the brittle rocks at shallower levels.

Data courtesy of Institute of Geological & Nuclear Sciences

ever, shallow earthquakes occur along all of the plate boundary, and reflect the widespread release of strain in the colder, brittle rocks of the shallow crust.

Given this picture of the collision that is taking place underneath New Zealand today, we may next wonder what effect plate tectonics has had on the New Zealand crust since collision began about 25 million years ago. Geologists have found the answers to this question in the rocks under the ocean floors – the basaltic crust has, in a sense, 'tape-recorded' the movement of the plates. Samples collected by drilling into the sea floor have confirmed that the basalt making up the oceanic crust becomes older the further away it is on either side of a mid-ocean ridge. More importantly, the rocks record changes that have taken place in the Earth's magnetic field.

Reversals of the north and south poles of the Earth's magnetic field happen every few million years, probably caused by changes within the core of the planet. At present, a compass needle will point to the North Pole, and this present magnetic field is referred to as 'normal'. However, during a 'reversal' a compass would point to the South Pole. Like a compass needle, the magnetic minerals in molten rock take up the Earth's magnetic field, and, as cooling takes place, the magnetic field at that time becomes permanently recorded in the rock.

Geologists have discovered that on either side of a mid-ocean ridge the crust consists of alternating bands of rock with opposite magnetism – either 'normal' or 'reverse'. Normal bands occur in oceanic crust that cooled when the Earth's magnetic field was the same as today; bands with reversed magnetism formed when the field was the other way round. This information, combined with radiometric dating, which determines when these rocks were formed, enables geologists to compile a precise record of magnetic reversals over the last few hundred million years. Because molten rock is continually upwelling at a mid-ocean ridge, the oceanic crust that is produced acts as a perfect recording device. As molten rock cools at the ridge, it takes up the magnetic field at that time, and is then conveyed away. The result is a striped pattern of magnetic anomalies, with a mirror-image pattern on either side of the ridge. Thus sea-floor spreading provides an extraordinary graphic record of the Earth's magnetic history.

It has thus been possible to chart the changing geography of land and ocean – in effect winding back sea-floor spreading and restoring continental crust back to former positions. When many different sections of oceanic crust are, on paper, moved in reverse and an ocean is closed up, geologists observe the movement is rotational about a point or pole. The pole of rotation for the Pacific Plate today is located about 2000 kilometres south-south-east of New Zealand (see figure on page 40).

When the south-west Pacific is closed up to the position of 25 million years ago, the various pieces of New Zealand end up in an unusual configuration (see opposite). It was around this time that the Pacific and Australian Plates began to push into one another across the New Zealand region. Where the plates

▼ With increasing distance from a mid-ocean ridge, oceanic crust and the overlying sediments become progressively older. The age of the sediments is known from the micro-organisms they contain, and radiometric dating gives the age of the oceanic crust. Periods of normal and reverse magnetism are 'logged' within oceanic crust as it is generated at a mid-ocean ridge, providing a graphic record of sea-floor spreading.

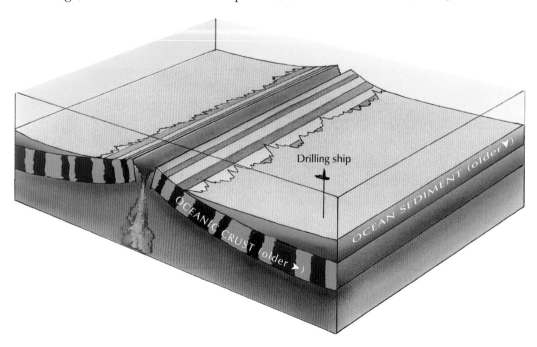

The New Zealand jigsaw

Collision between the Pacific and Australian plates during the last 25 million years has squeezed and rotated the various pieces of New Zealand into a quite different amalgamation. In the simplified diagrams below, the Arc ophiolites ——— and the Esk Head Mélange ——— are key geological markers that illustrate progressive fault movement. Parts of New Zealand's outline are included as a reference for separate crustal sectors.

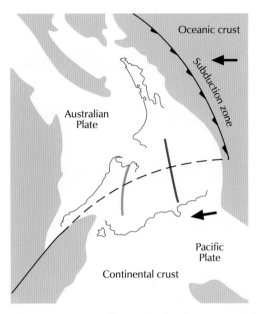

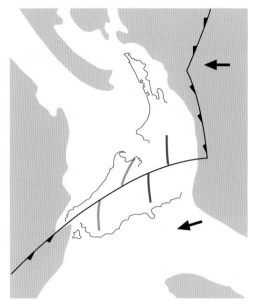

(i) 25 million years ago. As the Pacific Plate began to collide with the Australian Plate, a subduction zone developed to the northeast of New Zealand. Continental crust prevented the subduction zone from extending further south through the New Zealand region. The crust then began to compress and fracture, eventually forming a major transform fault, the Alpine Fault.

(ii) 10 million years ago. Movement on the Alpine Fault was tearing the New Zealand crust apart, displacing the rocks on either side. The Pacific Plate was advancing west and all of the New Zealand region was being compressed.

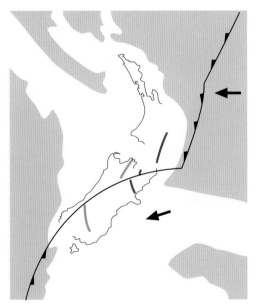

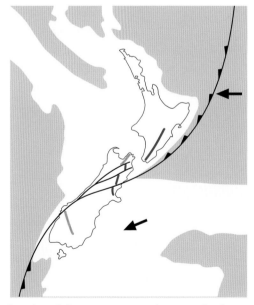

(iii) 5 million years ago. The continental crust had narrowed in width as it crumpled up. Mountains were being pushed up rapidly. Continental crust east of the Alpine Fault was being rotated anticlockwise and rammed into the Australian Plate, pushing up the Southern Alps. As the subduction zone extended further south, new faults began to splinter off the Alpine Fault.

(iv) Today. The subduction zone east of New Zealand now extends south almost as far as Kaikoura. The crust remains under immense compression as sea-floor spreading continues.

were in collision to the north, subduction took place, but across the New Zealand region subduction was blocked by continental crust. The New Zealand crust then began to compress and break up. At first, numerous small faults probably formed, eventually linking up to form the Alpine Fault. As the Pacific Plate rotated anticlockwise in relation to the Australian Plate, lateral movement began along the fault and the offset of the geological strata began.

▲ The Alpine Fault and the faults that branch from its northern end are known collectively as the Alpine Fault System. Hot springs ● are found along the faults where water makes its way to the surface from elevated temperatures at depth.

As movement continued, the northern edge of the continental crust riding on the Pacific Plate was carried progressively southward. This allowed the subduction zone to extend further south. As a result, new faults began to splinter off the northern end of the Alpine Fault. Over time, the major movement between the two plates in the northern South Island shifted south from the Wairau section of the fault on to parallel faults further south in the Marlborough area. Today, most of the plate movement is along the Hope Fault, which extends out to sea just north of Kaikoura (see above).

The Kaikoura Orogeny – uplift of the Southern Alps

While movement has been occurring on the Alpine Fault, the collision between the Pacific and Australian Plates has also been compressing the New Zealand crust, squeezing it tighter and pushing it up above sea level. All the mountain ranges and hilly ground, from the Fiordland mountains to the Ruakumara Range of East Cape, are the result of this compression. This collision and mountain-building event, which began about 25 million years ago, has been dubbed the Kaikoura Orogeny. It is in the Kaikoura region that the movement between the plates changes from ocean floor subduction to continental collision. As a result, blocks of land uplifted along faults form high mountains right at the coast, and close inshore the sea floor dives to depths of over 1,000 metres. It is this deep water that has put Kaikoura on the world map as a place to sight whales. After feeding on squid

▲ The Seaward Kaikoura Mountains rise to over 2600 metres just 12 kilometres from the sea. The Hope Fault, which marks the boundary between the Pacific and Australian Plates, runs along the foot of the Kaikoura Mountains, leaving the coast near the right of the picture.

in the deep canyon off Kaikoura, sperm whales surface and spout very close to land.

The mountain-building episode of recent times may have been named after Kaikoura, but the area of greatest uplift has been along the Southern Alps. Although these mountains probably began rising early in the Kaikoura Orogeny, geologists believe that their uplift has accelerated in just the last five million years, and is continuing today as fast as at any time in the past.

A cross-section through the South Island shows us how the Southern Alps are being pushed up. Near the surface the Alpine Fault dips eastward at about 55 degrees. The rock immediately west of the fault consists of granite and altered sedimentary rocks of the Australian Plate, while schist occurs on the east side. On a geological map (see page 16) schist rocks form a narrow band that runs the length of the fault.

▼ Cross-section drawn to scale through the South Island along a line between Mt Cook and Timaru. The Australian and Pacific Plates are being pushed into one another along the Alpine Fault. The continental crust of the Pacific Plate is being forced upwards along the fault. Recorded earthquake centres (o) are where the strain is being released by sudden movements.

Based on data in Rayners, 1987; Stern, 1995; Kleffmann et al., 1998

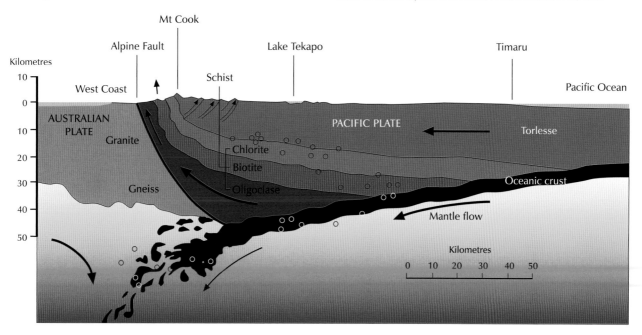

Adjacent to the fault are the most highly metamorphosed rocks – oligoclase and garnet schists. These grade eastward into biotite and chlorite schists and, finally, to Torlesse greywackes at the Main Divide. The schist closest to the fault has come from much greater depth than the rocks to the east. Apparently the schist strata have been thrust upward along the Alpine Fault, bringing them to the surface at a steep angle. On the ground we are thus presented with a cross-sectional view of metamorphic rocks from different depths. Below the surface to the east it is likely that zones of schist extend deep beneath the Torlesse rocks.

At depth the Torlesse rocks and underlying schist would be expected to rest on oceanic crust. Seismic studies, which record the shock waves from explosive charges, indicate that the Alpine Fault is a curved structure extending to a depth of at least 30 kilometres, and that continental crust rests on a layer of oceanic crust at a depth of 40–45 kilometres just east of the Southern Alps.

Gravity surveys across New Zealand indicate that the continental crust reaches a maximum thickness of about 45 kilometres. Adding to this picture is the distribution of earthquake centres. After an earthquake, the readings of seismographs stationed around the country are compiled to give the exact location and depth of the shock wave. The data collected since recordings began show that quakes are concentrated in several zones beneath the South Island (see cross-section previous page). Earthquakes at shallow levels may be where the Torlesse rocks are more brittle, while deeper quakes are thought to be where oceanic crust of the Pacific Plate is descending into the mantle. Surprisingly, there is a lack of earthquakes along the Alpine Fault, and the reasons for this remain the subject of intense debate.

By calculating the rates of sea-floor spreading, geologists estimate that the collision of the Pacific and Australian Plates has compressed the New Zealand crust about 60 kilometres in just the last five million years, and probably by a similar amount previously. This squeezing has gradually thickened the crust and inexorably pushed up the Southern Alps. Overall, the crust under the Southern Alps has been pushed up more than 20,000 metres — about six times higher than the mountains that exist today. How could this much uplift have taken place, and where has all the material gone?

Part of the answer lies in the Alpine Fault. While much of the movement along the fault is horizontal (averaging 20–30 millimetres per year), the east side is also being pushed up relative to the west side by as much as 10 millimetres a year. These rates have been determined by studying features such as river terraces and stream channels that have been offset by the fault (see below). This movement is not slow and continuous, but takes place intermittently. A rupture occurs when strain that has built up over a long period of time reaches breaking point. The sudden release of that strain causes a massive earthquake, and the land on either side may move horizontally as much as eight metres.

Such events have occurred about once every 100 to 300 years. The scientific evidence comes from the

One kilometre north of the Haast River an offset creek illustrates movement on the Alpine Fault.

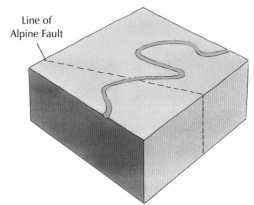

▲ Four thousand years ago a river terrace was formed – the age of this terrace is known from radiocarbon dates of wood samples. A creek then cut its channel across the terrace.

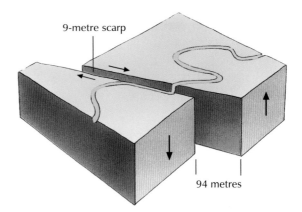

▲ Today we can see the effects of repeated movements along the Alpine Fault. The land on the east side of the fault has been raised nine metres and the course of the creek has been offset 94 metres. This gives an annual lateral movement rate of 23.5 millimetres.

Based on Cooper & Norris, 1995

radiocarbon dating of plant material found in trenches dug across the fault, and the age of plant material recovered from landslides caused by earthquakes. Corroborating evidence comes from growth-ring studies of long-lived trees, which can trace forest growth back as much as a thousand years. It seems that areas of forest not far from the Alpine Fault were disturbed or destroyed at specific times, and then suddenly succeeded by new forest. Major ruptures on the fault have thus been pinned down at about the years 1460, 1630 and 1717, the land being torn sideways by as much as eight metres each time. The 1717 rupture extended along the fault from Milford northward for 375 kilometres.

New Zealand's deep earthquakes are concentrated along the subduction zones that dip under the North Island and Fiordland (see page 41). One would also expect earthquakes along the connecting Alpine Fault, where the plates are being pushed past one another, but few deep earthquakes have been recorded there. The last major rupture (1717) was nearly 300 years ago, well beyond the 200-year average. This suggests we are overdue for a major earthquake on the Alpine Fault, and many scientists maintain that a build-up of strain on the fault will culminate in a catastrophic earthquake in the not-too-distant future.

There is, however, another school of thought. Geologists have found that only about 75 per cent of the movement between the plates is taken up on the Alpine Fault. It seems that the remaining 25 per cent of plate motion is distributed through the rock lying to the east of the fault. How this works is not well understood, but perhaps deep down, where temperatures are high and rock is more 'plastic', the thin layers of metamorphic minerals are slipping over one another throughout the rock mass. It is thought that ongoing metamorphism at depth is releasing small amounts of water in the rock, making the rock more ductile (able to be drawn out, like toffee).

Perhaps this ductile behaviour, or 'creep' as it often called, also applies to much of the movement along the Alpine Fault at depth where schist is being sheared against more rigid rocks of the Australian Plate. Could the plates just be creeping quietly past one another, the rocks deforming both at the fault, and within the schist to the east, without any violent rupture? This would explain the lack of large earthquakes at depth within the schist and along the

Alpine Fault itself. Perhaps deep earthquakes are not typical along the Alpine Fault. Ruptures at the surface may simply be confined to the strata at shallow levels – possibly in the upper five kilometres of the crust.

In recent years, satellite technology has begun a new revolution in studying the behaviour of the Earth's crust. With GPS (Global Positioning System) it is now possible to measure points on the ground within millimetres and, over time, to observe deformation of the crust. An experiment in the South Island conducted between 1994 and 1998 (see below) clearly shows how the ground in the eastern South Island is moving at up to 30 millimetres per year in a southwest direction relative to the land west of the Alpine Fault. The observed direction of movement of points on the Pacific Plate is rotational around the pole of rotation (see figure on page 40) positioned far to the south and east of New Zealand.

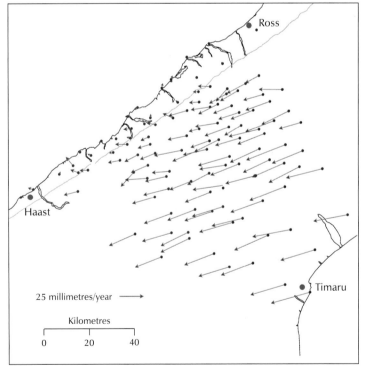

▲ Using satellite measurements (GPS) scientists have determined the positions of 115 ground stations at intervals between 1994 and 1998. The diagram shows the direction and amount of movement of the stations relative to a fixed reference west of the Alpine Fault. The arrow lengths express the rate of movement compared with the 25 millimetre per year arrow in the key. The data shows how the Pacific Plate is moving obliquely into the Australian Plate. Many points on the ground are moving at a rate of around 30 millimetres a year.

Map courtesy of John Beavan, Institute of Geological & Nuclear Sciences

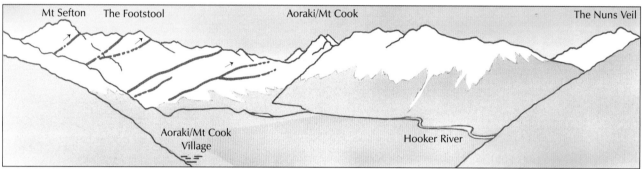

Mt Sefton   The Footstool   Aoraki/Mt Cook   The Nuns Veil

Aoraki/Mt Cook
Village

Hooker River

Bivuoac Fault

Semi-schist

Greywacke   Black Bob Fault

▲ Two faults within the Main Divide Fault Zone are seen here above the Hooker Valley in Mount Cook National Park. Semi-schist rocks (upper right) have been pushed up from depth along these faults, and now sit on top of greywacke rocks (lower left). The semi-schist rocks, once subjected to higher pressures and temperatures deeper in the crust, are quite different in character to the greywacke rocks. Broken and crushed rock can be seen across a wide area along the faults.

Photo and geology courtesy of Simon Cox

Within about 20 kilometres of the Alpine Fault the ground movement rapidly decreases, graphically indicating the build-up of elastic strain close to the plate boundary. The rock under the South Island is simply absorbing the strain in an elastic manner rather like compressing foam rubber against a solid object. Another analogy is a slow-motion automobile crash test – the front of the car stops abruptly against the wall while the back end decelerates forward.

Common sense tells us that, as a relatively brittle material, rock cannot continue to absorb strain like a block of rubber. Sooner or later something must give – the strain will be released within the crust by the sudden rupture of a fault. That rupture may not necessarily be on the Alpine Fault. There are numerous smaller structures to the east, such as the Porters Pass Fault in inland Canterbury or the Ostler Fault south of Lake Pukaki, that could generate a large earthquake. There are also back-thrusts in the Southern Alps that are angled back towards the Alpine Fault. The best example is the Main Divide Fault Zone,

which runs through Mount Cook National Park close to the Main Divide. This zone consists of short fault segments, along which the west side is thrust up over the east. This back-thrust faulting is a secondary effect of crustal collision and contributes to uplift of the Southern Alps.

▲ The track to Mt Sebastopol overlooks Aoraki/Mount Cook Village, with the Main Divide spanning the horizon as far as Aoraki/Mt Cook. The key at left identifies the main features, including segments of the Main Divide Fault Zone and the direction of fault movement.

Fault traces from Cox & Findlay, 1995

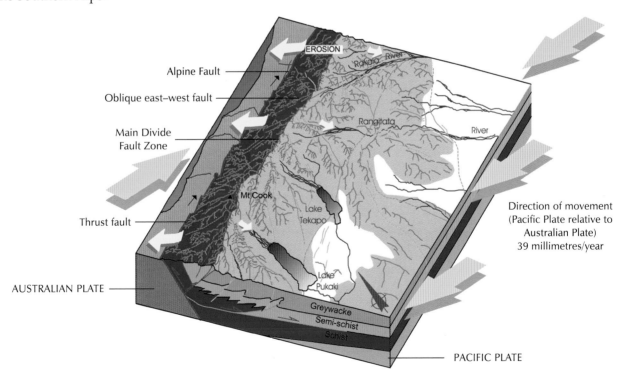

▲ Numerous active faults could trigger a major earthquake in the South Island. The main structure is the Alpine Fault. In its central section it is segmented by oblique east–west faults (lateral movement) and thrust faults (vertical movement). Running parallel to the Alpine Fault are back-thrust faults such as the Main Divide Fault Zone, along which collision has thrust the west side up over the east. Many other faults are scattered throughout the Southern Alps further east.

Diagram courtesy of Simon Cox, Institute of Geological & Nuclear Sciences

Uplift of the Southern Alps – an action replay

In these diagrams the crust has been stretched out to its former extent along a line across the South Island.
Then, keeping the volumes of rock constant, the Pacific Plate has been compressed into the Australian Plate.
The result is like an action replay of uplift along the Alpine Fault. Timaru and West Coast act as reference points.
The effects of erosion are omitted to show the total uplift. Measurements are in kilometres.

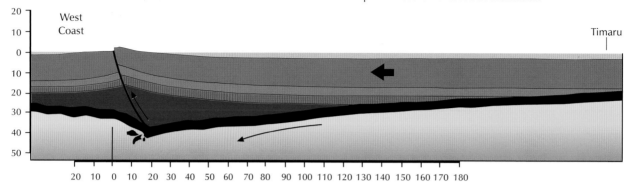

▲ From 25 to 15 million years ago most of New Zealand was still covered by ocean. Areas of land projecting above the sea were mainly Torlesse rocks. As the plates began to collide, the New Zealand crust came under pressure and the Alpine Fault was formed. The Haast Schists were still well below the surface along the line of the fault.

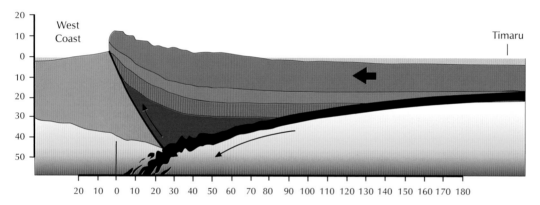

▲ Between 15 and 5 million years ago Gondwana rocks of the Australian Plate were carried north along the Alpine Fault and brought alongside Torlesse rocks. Chlorite-grade schist came to the surface. As the crust thickened under pressure, new areas of land were pushed above sea level and the Southern Alps were born.

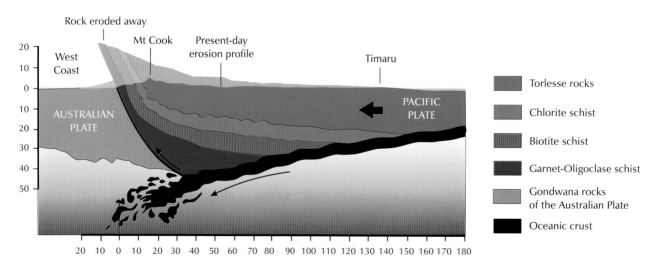

▲ From 5 million years ago to the present day the rate of uplift accelerated, pushing garnet-oligoclase schist above the surface. About 20 kilometres of uplift has taken place along the Alpine Fault.

With deformation and uplift occurring, not just along the Alpine Fault, but also within the schist and on the numerous faults to the east, we can begin to appreciate why the Southern Alps are among the fastest-rising mountains in the world. The Pacific Plate is literally being rammed up over the Australian Plate along a huge crack in the crust of the Earth. At the same time, the pressure is squeezing the rim of the Pacific Plate upwards. The resulting uplift rate of 10–20 millimetres per year is extreme compared with other mountain systems – in a million years it would mean more than 10 kilometres of uplift – but it has not always been this fast. Geologists believe it has accelerated in more recent time, and that the bulk of the Southern Alps have been pushed up in just the last few million years.

▲ If no erosion of the mountains had ever taken place, the Southern Alps would extend more than 20 kilometres into and beyond the stratosphere.

When we compare the total amount of uplift with the modern skyline of the mountains today, we can begin to appreciate the massive amount of erosion that must have taken place. The Southern Alps have been wearing down almost as fast as they have been rising. What remains of them today is tiny compared with the total amount of rock uplifted and eroded away. Something like half a million cubic kilometres of rock is missing (see the diagram opposite and the photograph above). How this material was removed, and where it has gone, is the final chapter in the story of the Southern Alps.

CHAPTER SIX

Erosion – shaping the Southern Alps from solid rock

W E TEND TO THINK of rock and mountains as solid and permanent, but stroll through any old New Zealand cemetery and you will see stone monuments with crumbling edges and illegible epitaphs, worn down by barely a century of exposure to rain, sun and wind. When it comes to mountains we have not just centuries but millions of years, and, given enough time, whole mountain ranges can disappear. In Chapter 5 we saw that a huge volume of rock has been eroded from the Southern Alps – in fact if there had been no erosion at all, the Southern Alps would be as much as six times higher than they are today.

Mountains start to disintegrate as soon as they begin to be uplifted. They literally come apart at the seams as pressure is released by the removal of overlying layers. Weaknesses in the rock, such as fractures, joints, faults, schistosity and bedding planes, begin to open up. This allows air and water to penetrate, and thus begins the process commonly known as weathering.

Weathering – how mountains begin to fall apart

Weathering may be mechanical, where rock is simply broken into smaller fragments, or it may be caused chemically, where oxygen and water combine with certain minerals to form soft weathering products such as oxides and clays. Over time, chemical weathering can severely weaken the mineral structure of rock, especially in warm, humid conditions. Minerals containing iron are particularly susceptible – the iron oxidises (rusts), staining rock surfaces yellow, red and brown. We see this type of weathering in outcrops of Torlesse well east of the Main Divide, where rock faces are less affected by run-off

Torlesse rocks, weathered to a rusty-brown colour, are exposed in the road cutting on the steep climb over Porters Pass. They consist of alternating beds of sandstone and mudstone that are steeply inclined and deformed.

and by the mechanical effects typical of higher mountain slopes.

However, across most of the Southern Alps chemical action is slow, because the atmosphere is cool and the main constituents of the Torlesse rocks – quartz in particular – are resistant to chemical alteration. Mechanical weathering is much more important. Heating in the sun by day, followed by cooling at night, expands and contracts rock, opening up small cracks. Water can then penetrate and, as it expands 4 per cent in volume when it freezes, the walls of cracks are pushed apart. With repeated freeze and thaw night after night, fragments of rock are broken off. Over time, these rock fragments accumulate below the mountain faces, forming steep slopes of loose scree.

Ice formation can also lift and move fragments of rock – a process known as 'frost heave'. As soil begins to freeze, crystals of ice grow vertically, raising

▲ These rock fragments on the Mt Sebastopol track, Mt Cook National Park, have been lifted several centimetres by 'needle ice'. This phenomenon occurs when the top of a soil layer begins to freeze. The needles grow vertically as moisture permeates upwards within the soil.

particles up off the surface. When the ice melts, the fragments inevitably move slightly downslope. In alpine regions, working unnoticed night after night, this process can move considerable amounts of rock debris downhill. Frost heave is thus as much an agent of transport as one of weathering. It is nature's quiet, first step in transporting material downslope and out of the mountains.

The ultimate goal of erosion is to transport the products of weathering out into the sea. Rivers and glaciers are the most obvious vehicles of erosion and sediment transport, but landslides play a fundamental part in demolishing the highest ground in the Southern Alps. Of special interest are rock avalanches, which are spectacular and dramatic in their effect on the landscape.

Rock avalanches – mountain demolition

Shortly after midnight on 14 December 1991, climbers camped at Plateau Hut beneath Mt Cook heard a loud rumble, which then became a roar like a raging river. Looking up towards the East Face in the dark, they saw orange sparks from rocks colliding with one another, and realised that a massive chunk of the mountain was tumbling down.

The noise continued for hours, and when dawn arrived the climbers viewed a scene of devastation. Jumbled rock and grey dust blanketed a huge area, and a large section of the mountain was missing from below the summit of Mt Cook. The debris had passed only 300 metres from where the climbers were camped, coating the hut with dust.

▲ These before (inset) and after views of Mt Cook show the section of mountain that parted from the summit. The figure of a climber at far left below provides scale.

Photo (after) M. J. McSaveney

This was New Zealand's most spectacular rock avalanche in historical times, all the more sensational because it involved the nation's most important icon, its highest peak. It lowered the summit of Mt Cook by 10 metres, to 3754 metres, and 14 million cubic metres of rock and ice were shed from the top of the mountain. During the collapse, even larger volumes of snow and ice were picked up downslope, giving the rock avalanche much greater size. It descended as a turbulent mass with fluid-like behaviour, reaching speeds of about 300 kilometres an hour and acquiring enough momentum to reach the far side of the

Tasman Glacier, six kilometres from the mountain summit. A deposit of rock and ice debris, much of which had been broken up into fine fragments during its fierce descent, was left strewn across the full width of the valley. In total the deposit was estimated to have a volume of around 55 million cubic metres.

The shock waves from this rock avalanche were big enough to register on a number of seismographs such as the one at Erewhon, 58 kilometres east of Mt Cook. The main collapse was recorded for just over a minute, with smaller rock falls continuing for hours afterwards. The shock waves recorded at Erewhon, Twizel and other stations clearly pointed to Mt Cook as the source of the seismic signature.

▲ The Mt Cook rock avalanche of December 1991 spread a trail of dark-coloured debris for six kilometres from the East Face, down the Hochstetter Icefall and right across the Tasman Glacier.
Photo T. J. Chinn

▼ The seismic signature of the avalanche recorded at Erewhon.

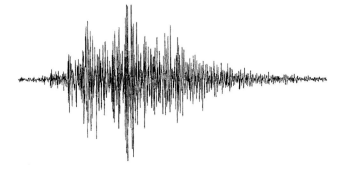

Earthquakes are the usual trigger of rock avalanches, but the Mt Cook event was an exception – seismographs did not pick up any earthquake beforehand. It seems that the collapse was simply due to the general weakness of the rock below the summit. When the East Face was inspected by geologists in a helicopter, they were surprised at the shattered condition of the rock, and likened the peak of Mt Cook to Weet-Bix. Progressive erosion by glacial ice at the bottom of the face may have been enough to destabilise the weak summit until it collapsed under its own weight. Whatever the trigger, it is just a matter of time before the same thing happens again.

Another rock avalanche of recent times occurred at Mt Fletcher, about 35 kilometres north-east of Mt Cook. In May and September 1992, parts of the east face of the mountain collapsed. Both rock avalanches charged down the Maud Glacier and into the lake at the foot of the glacier, leaving a deposit of dark grey rubble over the snow and ice. When the rock avalanche hit the water, a number of icebergs were washed out by a huge wave and left high and dry at the far end of the lake.

▼ After rock avalanches in September 1992 the normally snow-covered face of Mt Fletcher is left scarred and dirty. Twice that year, rock avalanches flowed down the mountain (as arrowed) and on down the Maud Glacier into the lake at the glacier terminus, covering a large area with debris and dust. Mt Fletcher lies in the upper catchment of the Godley River, which flows into Lake Tekapo 35 kilometres to the south.
Photo D. L. Homer

Inset: These icebergs at the far end of the lake were washed seven metres above the lake level by the flood wave after the avalanche.
Photo M. J. McSaveney

As with the Mt Cook event, there was no earthquake to trigger the Mt Fletcher collapses, and the rock forming Mt Fletcher is also weak and crumbly. Geologists said that if the summit of Mt Cook was like Weet-Bix, then Mt Fletcher could be compared with cornflakes. Both these mountains lie along the Main Divide Fault Zone. The numerous faults that make up this zone are marked by wide areas of shattered and pulverised rock. Mt Cook and Mt Fletcher are two casualties along this tract of unstable rock.

The Southern Alps are riddled with faults but, unlike the faults of the Main Divide Fault Zone, most are ancient, the legacy of earlier crustal collision. Movement along them may have ceased long ago, but they nevertheless form lines of weakness, with the Torlesse rocks often fractured for some distance on either side. As a result, mountain collapses are common in the Southern Alps. In the ranges between Arthur's Pass and Mt Cook some 46 rock avalanche deposits greater than a million cubic metres in volume, have been identified. Most of these were bigger than the Mt Cook rock avalanche, and some were

several hundred million cubic metres (up to 50 times bigger than the Mt Cook collapse). The largest in historical times, aptly named the Falling Mountain rock avalanche, was triggered by the 1929 Arthur's Pass earthquake, a 60-million-cubic-metre chunk of the mountain collapsing into the valley near Tarahuna Pass.

Also in Arthur's Pass National Park, but more familiar to the public because it is crossed by the highway, is the rock-avalanche deposit at the spectacular Otira Viaduct, a 440-metre bridge completed in 1999 to maintain State Highway 73 through Arthur's Pass National Park. Before the viaduct was built, cars wound their way over a steep, tortuous section of the highway known as the Zig Zag. This was probably New Zealand's most hazardous alpine road route, and keeping it open was difficult and expensive. Periodically the edge of the road would collapse into the Otira Gorge, sometimes requiring a major

▼ The tortuous Zig Zag route on State Highway 73, seen here before the Otira Viaduct was built, sits precariously at the edge of a steep unstable scree slope.
Photo D. L. Homer

▲ The Otira Viaduct now bypasses a steep unstable scree slope. The previous Zig Zag route crossed the scree at the upper right of the picture. The huge block in the riverbed has fallen from the ridge crest 700 metres above and is too large for the river to move downstream. It is one of a number of kaitiaki (sacred stones) that mark the Māori route through this pass.

realignment of the highway. Eventually it became impossible to shift the road any further upslope. Not only was the material under the road unstable, but motorists on the Zig Zag also faced falling rock debris from above, including some boulders as large as cars. The cause of this problem is the greatly fractured mountain face above the road, some 700 metres above the Otira River. Given the scale of all these problems, a long-term solution was always going to be expensive. At a cost of $25 million the Otira Viaduct now bypasses the unstable scree slope and takes traffic away from the hazard of falling debris.

The steep scree slopes above the viaduct are part of a huge rock-avalanche deposit. This material accumulated during several separate collapses of the mountain face above the Otira Gorge. The last collapse was 2000 years ago, as determined by radiocarbon dating carried out on fragments of trees buried by the landslide. Debris from the rock avalanche filled the valley well above the level of the viaduct. Obviously no human-made structure would be safe in another

▲ The mountain face above the Otira Viaduct consists of highly fractured rock with near-vertical open joints that continue well down the face. Some areas are semi-detached and poised for the next earthquake. The light-coloured patch at upper right marks where a section of rock has parted from the face in recent years.

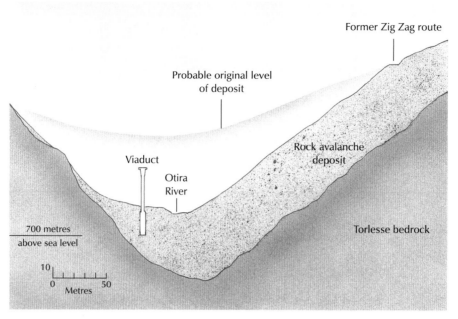

Former Zig Zag route

Probable original level
of deposit

Rock avalanche
deposit

Viaduct

Otira
River

Torlesse bedrock

700 metres
above sea level

10

0   50
Metres

▲ This section across the Otira Gorge shows the viaduct in the context of the rock avalanches that have dammed the Otira River in the past. Avalanche debris extends to considerable depth: the viaduct pier foundations were excavated to a depth of 25 metres, and pilot holes went to 40 metres without reaching bedrock. Since the valley was filled with debris, the Otira River has been slowly cutting down towards its old bedrock channel, steepening and destabilising the toe of the deposit.

rock avalanche of that scale, and indeed the viaduct was not built to withstand such an event. Its real purpose is to avoid the hazard of rockfall and the threat of collapse of the road along the unstable Zig Zag route.

Glaciers – mountain excavators

Unlike rock avalanches, which perform only on rare occasions, glaciers are ceaseless workers. Although they only move very slowly, in the Southern Alps glaciers have been reshaping the landscape relentlessly for the last two million years, and today we can still see them at work along the Main Divide.

Most familiar to the public are the Franz Josef and Fox Glaciers (Kā Roimata o Hine Hukatere and Te Moeka o Tuawe) in South Westland, which are spectacular tourist attractions because they extend right down into the forest. However, there are literally hundreds of other glaciers along the Main Divide. They owe their existence to the high elevation of the Southern Alps, which emerge more steeply from the sea than any other mountain chain on Earth. Unlike the Andes or the Rockies, which lie a considerable

distance inland, the highest peaks of the Southern Alps are only 20 kilometres from the coast and form a long barrier across the path of storm fronts approaching from across the Tasman Sea. Moisture-laden westerly winds then rise up the mountain slopes, cooling as they ascend to form clouds that dump large volumes of rain and snow on alpine areas. This leads to a dramatic increase in annual precipitation, from around three metres at the coast to as much as 13 metres on the slopes west of the Main Divide (see page 62). In some areas annual precipitation totals 16 metres, most of it falling as snow.

The development of glaciers is inevitable wherever this much snow accumulates. Under its own weight the snow packs down and air is squeezed out until blue glacial ice forms. We are all familiar with ice as a slippery material, and even a large body of ice will not rest still on a sloping surface. This is precisely what a glacier is – a large body of ice moving slowly downhill. The movement is almost imperceptible – stand next to a glacier and you probably won't see or hear anything happening.

The rate at which a glacier moves depends on a number of factors. New Zealand's biggest glacier, the Tasman Glacier in Mt Cook National Park, moves quite slowly at less than a metre a day because it occupies a wide, gently sloping valley. In contrast, the Franz Josef Glacier moves four metres per day where ice in the large névé at the top of the glacier is squeezed into a narrow, steep valley. The overall movement of the Franz Josef Glacier was shown

▲ The Southern Alps stand like a fortress wall in the path of the westerly air flow. This view from the Tasman Sea includes Mt Cook on the right and the Franz Josef Glacier to the left.

Photo D. L. Homer

▼ The Franz Josef Glacier is a popular stop for visitors, but care is needed close to the ice. In January 1994 heavy rain led to a spectacular collapse at the glacier, washing blocks of ice far downstream.

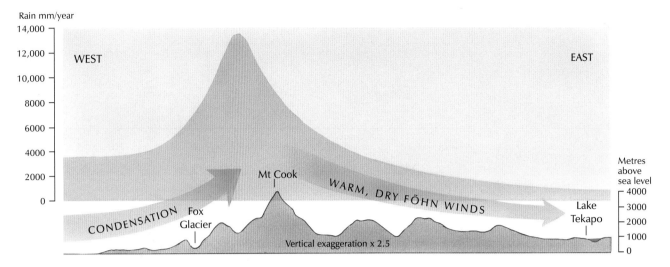

Rain mm/year

14,000

12,000

10,000

8000

6000

4000

2000

0

WEST

EAST

Mt Cook

WARM, DRY FÖHN WINDS

CONDENSATION

Fox
Glacier

Lake
Tekapo

Metres
above
sea level

4000

3000

2000

1000

0

Vertical exaggeration x 2.5

▲ Across the Southern Alps, from west to east, annual precipitation climbs rapidly to a maximum of around 13 metres west of the Main Divide. The rainfall graph given here is compiled from gauges in the vicinity of the section line (shown for reference). After the release of moisture in the west, the airflow descends eastward to become a warm, dry wind, giving Canterbury its renowned 'nor'wester'.

Rainfall data from Henderson & Thompson, 1999

by the wreckage of a plane that appeared at the bottom of the glacier in 1949, six years after it had crashed on the ice about four kilometres up from the terminus.

Although a glacier only moves slowly, over time it can move massive volumes of debris. Acting like a giant conveyor, it will soon carry away rock that falls from the valley sides and onto its surface. For example, within the next hundred years all the debris that

fell onto the Tasman Glacier during the 1991 Mt Cook rock avalanche will reach the glacier terminus 11 kilometres down-valley.

However, most of the debris transported by a glacier is carried not on its surface but within the ice itself. As the ice grinds over the bedrock, pieces of rock are continually broken off, scraped up and carried away. Most of this sediment load remains hidden within the glacier until it is exposed by rapid melting at the terminus. Over thousands of years the scouring action of a glacier will excavate a huge

▼ The highway to Mt Cook passes large mounds of debris (moraines) after crossing Freds Stream. Carried here by the Tasman Glacier about 10,000 years ago, the debris occupied hollows in the ice. Then, when the ice melted, it was left behind as mounds, forming this 'knob-and-kettle' topography.

▲ The terminal moraine known as the Waiho Loop was left by a minor advance in the Franz Josef Glacier 12,000 years ago, before the glacier continued retreating to its present-day position. The moraine forms a tree-covered semicircle of debris 80 metres high and 5 kilometres long. Forest growing on the moraine was spared pastoral clearing by early farmers. The township of Franz Josef Glacier/Waiau can be seen at far right.

Photo D. L. Homer

amount of rock, deepening its valley and carving it into a distinctive U-shape. Anything from blocks the size of buildings to finely ground 'rock flour' is swept along. Some of this material may be deposited along the sides of the glacier in elongated mounds called lateral moraines, but most is delivered to the terminus, from where it is either washed downstream or deposited in a heap called a terminal moraine.

Wherever moraines survive later erosion, they provide geologists with valuable clues about the extent of glaciers in the past, and almost anywhere in the foothills of the Southern Alps the trained eye can spot a host of glacial landforms. In New Zealand, studies on moraines and other ancient glacial features have revealed a past history of glaciers that were huge compared with those of today. Clearly, New Zealand was caught in the grip of the great Ice Ages that affected much of the Earth during the last two million years.

This last period in the history of the Southern Alps was a time of global cooling, with average temperatures dropping as much as 4.5°C. Glaciers worldwide advanced dramatically, and broad ice sheets formed in the Northern Hemisphere. In New Zealand, glaciers expanded out from the Southern Alps, carving huge valleys east of the Divide. Later, some of these valleys filled with water to become Lakes Tekapo, Pukaki, Ohau, Hawea, Wanaka, Wakatipu, Te Anau and Manapouri.

The Ice Ages span the period of geological time known as the Quaternary. Throughout that time, fluctuating climate caused glaciers to advance and retreat repeatedly. There is little record of New Zealand's early glaciers – the features they produced have been eroded or were overrun and destroyed by later advances. The record starts to become clear from 250,000 years ago, and there is evidence that glaciers extended great distances down many of the main valleys. For example, in the upper Waimakariri, near Arthur's Pass, the ice was over 1000 metres thick and at times extended east through the foothills of the Southern Alps to the edge of the Canterbury Plains.

There were four major cycles of advance and retreat in the South Island during the last 250,000 years.

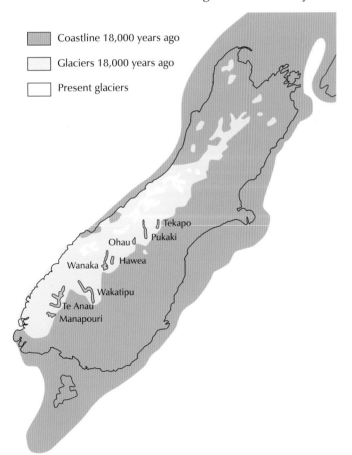

Coastline 18,000 years ago

Glaciers 18,000 years ago

Present glaciers

▲ During the Late Otiran Glaciation 18,000 years ago, large glaciers covered all of the Southern Alps and the coastline was a long way further out.

Redrawn from Suggate, 1990

The most recent of these, and the best understood, has been called the Otira Glaciation, which reached its maximum 18,000 years ago. At this time the sea level was as much as 130 metres lower than today because so much water was locked up in the great ice sheets of the Northern Hemisphere. As a result, parts of the New Zealand shoreline were many tens of kilometres out from the present-day position. Large tongues of glacial ice extended well beyond the present coastline of the West Coast.

About 14,000 years ago the Quaternary glaciers

began to melt away, or 'retreat', as temperatures rose rapidly. As the ice thawed, large melt-water lakes often formed in front of glaciers and behind end moraines far down-valley. Eventually the ice retreated to its present-day extent, leaving glaciers that are tiny in comparison to their huge ancestors.

Less snow falls on the Southern Alps today than in the past. Precipitation still remains high overall, but mostly it is in the form of rain. Thus today, erosion, and the transport of mountain debris, takes place largely through run-off and the action of rivers.

▲ The Southern Alps are seen here from Mt Mary in the Mackenzie Basin. Mt Cook is on the far right. The mist, which hides Lake Pukaki from view, fills the valley just as glacial ice would have 18,000 years ago, and provides a hint as to how the place would have looked. Near-horizontal glacial terraces on the far side of the valley, just above the mist, were formed earlier when the ice was at a higher level.

▼ The Southern Alps, with Mt Cook in the distance, are reflected in Lake Pukaki, a natural lake scoured out during the Ice Ages. The level of the lake is now controlled for hydro-electric power generation. Glacial moraines form all the lake shores.

▲ In the upper Waimakariri Valley below Mt Binser a series of lake beaches can be seen cut like a flight of stairs into the foot of the Binser fan (centre of photo and running off to the right). These were formed during the life of the now-extinct 'Glacial Lake Speight' that filled the valley for a distance of 15 kilometres as the Waimakariri glacier melted near the end of the Otira glaciation. The lake became lowered intermittently, and the beaches were cut by wave action at each successive drop. These ancient lake shores can be seen a kilometre past Lake Sarah on the Craigieburn Road, off the main highway just south of Cass.

Rivers and debris flows – mountain transport

If we liken a glacier to an excavator digging a route through the mountains, then we could compare a river with an earth-moving truck, transporting rock waste to some distant site for landfill. The two work together – material is transferred from excavator to truck, glacier to river, at the glacier terminus. However, each truck-load of sediment must wait for its departure – river transport will only swing into action when the next big rainstorm arrives.

This is how rivers move sediment – not continuously like glaciers, but intermittently during periods of flood. On a fine day a river may look placid, but this belies its power and efficiency in moving massive amounts of material.

Mountain rivers

We can begin to appreciate the erosive power of rivers in the Southern Alps when we look at the rainfall statistics. While annual rainfall may be up to an amazing 13 metres on the slopes west of the Divide,

the data for single rainstorms is even more astonishing. In just a 24-hour period, gauges have been known to collect as much as a metre of rain. This much water spread across a vast area is bound to have dramatic effects. Streams and rivers become raging torrents, scouring material from their channels and picking up even the largest boulders. Although floods may last only a day or two, during that time huge amounts of sediment can be moved. Once the flow subsides, however, the sediment, beginning with the coarsest material, will start to settle to the bottom. Eventually everything – boulders, gravel, sand, and mud – will come to rest somewhere downstream. Over thousands of years repeated floods can, in effect, move whole mountain ranges into the sea. If the Southern Alps were not still rising rapidly, their lofty peaks would have disappeared long ago.

Evidence of the rapid erosion taking place in the Southern Alps can be seen in the glaciated valleys east of the Main Divide. For example, the once deeply glaciated valley of the upper Waimakariri River has been infilled with so much sediment since the ice retreated, it now has a wide, flat floor (see top right).

Debris flows

Other evidence of extreme run-off and rapid erosion are the many fan-shaped piles of debris seen along the sides of the main valleys of the Southern Alps. Known as alluvial fans, these structures can be very large and may contain a huge volume of material.

However, as indicated by the grass and scrub covering their surface, the streams that supply them with sediment are rarely active. It is not the frequency of the flow that is important on alluvial fans, but the power of the flow – the power of a spectacular phenomenon known as a 'debris flow'.

In April 1978, severe rainstorms over several days dumped more than 300 millimetres of rain at Mt Thomas on the western side of the Canterbury Plains. After locals raised the alarm, geologists hurried to the Bullock Creek area to witness the rare sight of

▲ The upper reaches of the Waimakariri River occupy a large glaciated valley. Since the ice retreated, a great thickness of sediment has accumulated in the valley, forming the wide riverbed.

▼ Four kilometres south of Cass, near Arthur's Pass, State Highway 73 (along tree line) climbs over a large alluvial fan that covers an area more than three kilometres wide between Lake Grasmere (foreground) and Lake Pearson (far left). The fan debris has come out of the narrow gully at the top of the fan. The build-up of material probably took place rapidly from about 10,000 years ago after the glaciers last retreated. Ribbonwood Stream is now cutting a course down into the fan. Smaller alluvial fans can be seen either side of the Ribbonwood fan.

▲ Debris flows at Mt Thomas in 1978.

Photo D. W. Lewis

debris flows in action. A dirty liquid that looked and flowed like motor oil was gushing out of a ravine at the head of the alluvial fan, carrying with it mud, sand, gravel and boulders. Surges would occur about every 15 minutes, the thick slurries sometimes racing down the channel at 20 kilometres per hour with a wavefront as high as three metres. Some of the thicker flows were more than three-quarters solid material, rather like dirty, runny concrete.

For three days the debris flows continued as the rain persisted. In the catchment area, large areas of weak, shattered Torlesse rock became saturated, and parts of the slope slid into the ravine, one after another. On the upper fan the slurries surged from the channel and washed out over the lower fan, travelling several kilometres downslope. By the time the event was over, nearly 200,000 cubic metres of gravel, sand and mud had been spread over the fan surface in a layer up to three metres thick. It would have

taken several thousand years to transport this much sediment by normal stream erosion.

This was not just a one-off. According to local accounts, debris flows have been occurring there every 10 to 20 years, and geologists believe they have been building out the fan for at least the last 20,000 years. Debris flows are now regarded as a major player in the erosion of the Southern Alps. In areas like Mt Thomas, where the rock has been weakened by faulting across a wide area, their effects are especially important.

We can see the past work of debris flows in the many alluvial fans found throughout the Southern Alps. Often formed where a side valley enters a major river valley, alluvial fans are like stockpiles, where debris flows dump their load for a main river to dispatch later.

Where rivers meet the sea
Once the main rivers leave the Southern Alps they flow out across coastal plains, and geologists have uncovered an impressive story about the past effectiveness of rivers on both sides of the South Island to move sediment out of the mountains. At Haast on the West Coast, for example, rivers have built out a coastal plain 50 kilometres long and 10 kilometres wide in just the last 6000 years. The many beach

▼ Rivers draining the Southern Alps have deposited their sediment load to form coastal plains. The Haast coastal plain has formed where several rivers draining a wide area of the mountains converge along a short stretch of the coastline. The Canterbury Plains have been constructed mainly by the Waimakariri and Rakaia Rivers.

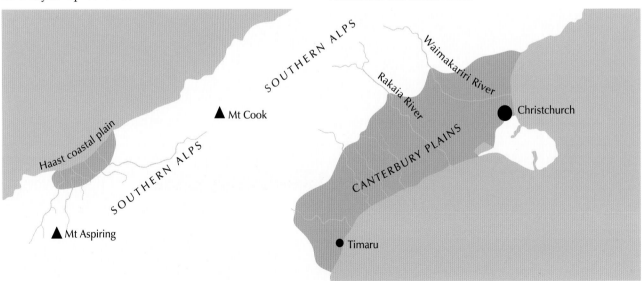

ridges running along this coast represent old shorelines, each one left behind when the coastline advanced seaward to create a new beach. Each advance of the coastline occurred when massive floods poured huge amounts of sediment into the sea, with waves and shore currents moving the sediment along the coast until a new beach line was formed.

▼ The coastal plain at Haast features a series of ancient beach ridges. Forest grows on these ridges, with swamps and lakes between them. The Okuru and Turnbull Rivers emerge at the coast in the distance.
Photo D. L. Homer

A number of rivers, including the Haast, Arawata, Waiatoto, Okuru and Turnbull, have worked together to build out this coastline in a short space of time. In total, the catchments of these rivers drain a 150-kilometre section of the Southern Alps, stretching from Mt Cook in the north to beyond Mt Aspiring in the south. In their journey out of the mountains, these rivers all converge on the coast at the Haast coastal plain, where, every year, they dump about 50 million tonnes of sediment into the Tasman Sea. Most of this is then pushed back onto the shore by the westerly swell.

To the east of the Southern Alps are the more extensive Canterbury Plains, a large flood plain about 160 kilometres long by 50 kilometres wide created by a number of braided rivers emerging from the Southern Alps. The largest of these rivers are the Rakaia and Waimakariri. Throughout the Quaternary age, the Canterbury rivers laid down countless layers of gravel, sand and mud across the flood plain, building out the coastline. A great thickness of river sediments accumulated – drill-holes west of Christchurch have found gravel and sand down to a depth of more than 500 metres.

During the Ice Ages the types of sediment deposited on the flood plain changed as the climate fluctuated. When the glaciers were in advance, thick layers of gravel and sand accumulated. Then, during periods of glacial retreat, the sea encroached on the land so that muddy marine sediments were laid down.

These alternating layers of fine and coarse sediment have given Christchurch its wonderful source of artesian water. West of the city, water from the Waimakariri seeps into the gravel layers beneath the

▼ Braided channels of the Waimakariri River carry sediment eroded from the Southern Alps to the Canterbury Plains. The term 'braided' refers to the tangled pattern of channels, in contrast to the winding course of a meandering river.
Photo D. L. Homer

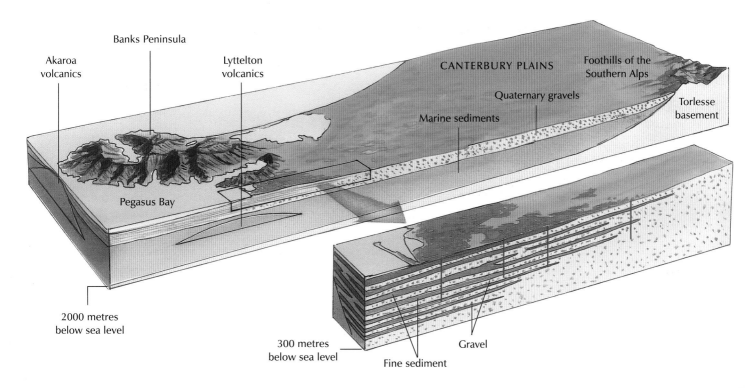

Akaroa
volcanics

Banks Peninsula

Lyttelton
volcanics

CANTERBURY PLAINS

Foothills of the
Southern Alps

Quaternary gravels

Marine sediments

Torlesse
basement

Pegasus Bay

2000 metres
below sea level

300 metres
below sea level

Gravel

Fine sediment

▲ The sedimentary layers that underlie the Canterbury Plains are illustrated in this simplified cross-section (vertical exaggeration x 5). Overlying the Torlesse basement are strata ranging in age from the Cretaceous to the present day. Layers of sand and gravel, deposited in glacial times, extend a long way east of the present coastline. The expanded central section (vertical exaggeration x 20) shows some of the detail that has been obtained from hundreds of Christchurch well holes, a few of which are shown (in blue) in the diagram. The layers of porous sand and gravel, sandwiched between impervious layers of fine sediment, are the aquifers that give Christchurch its artesian water supply.

Aquifer detail redrawn from Brown & Weber, 1992

riverbed. Moving slowly eastward, the ground water becomes trapped within the gravel layers by intervening layers of muddy sediment that are impervious to water. Thus the gravel layers (known as aquifers) act as water conduits that can be tapped into by putting down bores.

Before human settlement the Canterbury rivers were free to wander across their broad flood plain. During times of flood, rivers would burst their banks and inundate wide areas. The new settlement of Christchurch, built where channels of the Waimakariri River sometimes roamed across the plain, was prone to such floods in its early days. The town suffered its worst flood in February 1868, after which stopbanks were constructed to confine the river to one channel. Since the 'Great Flood' these protection measures have managed to keep the Waimakariri within its

channel. However, city planners are aware that a 'one-in-200-year flood' could overtop the stopbanks and flood Christchurch. As with all natural hazards, whether they be earthquakes, volcanoes or floods, if the event is big enough there is little we can do to safeguard ourselves.

While the Waimakariri may be confined to a single course, other Canterbury rivers burst their banks at odd intervals, flooding large areas of farmland with muddy water (see page 72). Such events are part of the natural process of flood-plain development. Sometimes a river in flood will cut a new course across its plain, the new channels filling with sand and gravel as the flow subsides. Then the muddy flood waters that may stretch far and wide slowly dry out, leaving new layers of soil across the surface of the land. Over time, with flood after flood, the plain gradually builds up layer by layer.

While the Canterbury Plains are New Zealand's largest alluvial plain, what we see today is only small in comparison with the much larger flood plain that existed during glacial times. Then, the sea level was so much lower that the rivers draining the glaciers flowed out to a coastline far to the east. Thus today the river sediments that underlie the Canterbury Plains extend beneath the sea a long way east of the present-day coastline. This is where a big chunk of the Southern Alps has been 'laid to rest'.

▲ In August 2000 the Selwyn River, on the Canterbury Plains near Christchurch, burst its banks and flooded a wide area of farmland. This is an event that has doubtless been repeated many times as the plains have been built, and serves to illustrate how the process is still continuing.

Photo Selwyn District Council

▼ During a period of flood, sediment is flushed out of the Waimakariri River mouth and into Pegasus Bay. The discharge of the river takes time to mix with the denser sea water, leaving a dirty halo projecting into the sea. During serious floods of the past, much of the plains seen here would have been awash.

Photo Don Scott, *The Press*

Today, when the Waimakariri and Rakaia Rivers are in flood, large volumes of sand and mud are washed out into the sea. Waves and currents then sort the sediment and redistribute it. Along the east coast of the South Island, ocean currents slowly transport sediment from south to north. However, within Pegasus Bay, in the lee of Banks Peninsula, the currents spiral around in a back-eddy, carrying sediment southward to form the South Brighton spit separating the Avon-Heathcote Estuary from the sea. Outflows from the estuary are in constant battle with the coastal currents, which are always trying build the spit across the mouth of the estuary. From time to time the channel must be dredged to keep it in the same position.

As the sea drags sand along the coast, wave action then pushes some of it onto shore along the beaches of Brighton and North Beach. Wind off the sea then blows some of the sand into dunes behind the beaches. All the grey sand along the Pegasus Bay coastline is, in essence, the ground-up remains of the Southern Alps. Most of those remains, however, are nowhere to be seen, long since carried away by bottom currents across the floor of the ocean on either side of the country.

▼ Banks Peninsula, in the distance, interrupts the northerly drift of sediment along the Canterbury coast. The resulting back-eddy has built South Brighton spit southward across the mouth of the Avon-Heathcote Estuary. The beaches of Pegasus Bay are backed by a continuous line of sand dunes.

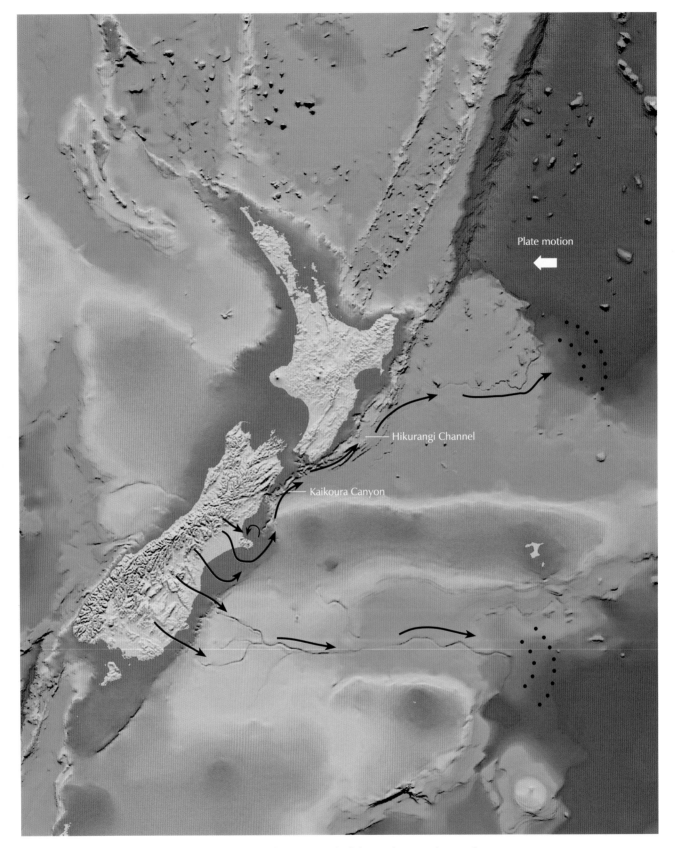

Plate motion

Hikurangi Channel

Kaikoura Canyon

The fate of sediment eroded from the Southern Alps

East of the South Island sediment eroded from the Southern Alps is carried northward by ocean currents. Some of the sediment from the Canterbury rivers is trapped by a back-eddy in Pegasus Bay, but most of it drifts north into the Kaikoura Canyon. Sediment is then carried an extraordinary distance along the Hikurangi Channel to fans at the edge of the Hikurangi Plateau far east of the North Island. Sediment discharged from the Waitaki and Clutha rivers travels along submarine canyons out to the Bounty Fan some 1000 kilometres east of the South Island.

Bathymetry courtesy of NIWA

It is an interesting quirk of fate that some of the sediment eroded from the Southern Alps is today being recycled once again. Transported along extensive submarine channels, it finally comes to rest on submarine fans far to the east of New Zealand. Sea-floor spreading is then pushing the submarine-fan sediment back into the subduction zone east of the North Island. Like a case of déjà vu, this is similar to the picture we have seen before, when Torlesse submarine-fan sediments first began arriving on the sea-floor conveyor 250 million years ago.

We could say that, in this story of the Southern Alps, we have come full circle to the place where we began. It has really been the story of a sand grain's journey, a journey that began on an ancient sea floor and has finished on the floor of the sea today. That journey has taken us on a tour of the 'rock cycle'

several times. First the sand grain was eroded from ancient granite mountains, and deposited on the sea floor. Then, sea-floor spreading carried it into a subduction zone, where it was buried, becoming part of the schist rocks of the region. Later, the remains of that sand grain were pushed back up as part of the Rangitata mountains. Upon subsequent erosion of those mountains, the cycle was repeated again, and today once more the cycle is being repeated with the erosion of the Southern Alps. As consistent and as cyclical as nature tends to be, we can be sure that one day in the distant future what remains of that original grain will be part of a completely new chain of mountains.

▼ Greywacke sand in Pegasus Bay – the ground-up remains of the Southern Alps.

Adamss, C. J. D., 1979. 'Age and origin of the Southern Alps', in R. I. Walcott & H. M. Cresswell (eds), *Origin of the Southern Alps*, Royal Society of New Zealand Bulletin 18: 73–78.

Adams, J., 1979. 'Vertical drag on the Alpine Fault, New Zealand', in R. I. Walcott & H. M. Cresswell (eds), *Origin of the Southern Alps*, Royal Society of New Zealand Bulletin 18: 47–54.

Allis, R. G., 1986. 'Mode of crustal shortening adjacent to the Alpine Fault, New Zealand', *Tectonics*, 5: 15–32.

Beaven, J., M. Moore, C. Pearson, M. Henderson, B. Parsons, S. Bourne, P. England, D. Walcott, G. Blick, D. Darby & K. Hodgkinson, 1999. 'Crustal deformation during 1994–1998 due to oblique continental collision in the central Southern Alps, New Zealand, and implications for the seismic potential of the Alpine fault', *Journal of Geophysical Research*, 104: 25, 233–25, 255.

Berryman, K. R. & S. Beanland, 1988. 'Ongoing deformation of New Zealand: rates of tectonic movement from geological evidence', *Transaction of the Institutuion of Professional Engineers New Zealand, General Section*, 15(1): 25–35.

Bradshaw, J. D., 1989. Cretaceous geotectonic patterns in the New Zealand region. *Tectonics*, 8: 803-820.

Brown, L. J. & J. H. Weber, 1992. *Geology of the Christchurch urban area. 1:25,000 scale*, Geological map + 104 pp, Institute of Geological & Nuclear Sciences Ltd, Lower Hutt.

Carter, L., R. M. Carter, I. N. McCave & J. Gamble, 1996. 'Regional sediment recycling in the abyssal Southwest Pacific Ocean', *Geology*, 24: 735–38.

Cooper, A. F. & R. J. Norris, 1995. 'Displacement on the Alpine Fault at Haast River, South Westland, New Zealand', *New Zealand Journal of Geology and Geophysics*, 38: 509–14.

Cox, S. C. & R. H. Findlay, 1995. 'The Main Divide Fault Zone and its role in the formation of the Southern Alps', *New Zealand Journal of Geology and Geophysics*, 38: 489–99.

Gage, M., 1958. 'Late Pleistocene glaciation of the Waimakariri valley, Canterbury, New Zealand', *New Zealand Journal of Geology and Geophysics*, 1: 103–22.

Grapes, R. M., 1995. 'Uplift and exhumation of Alpine schist, Southern Alps, New Zealand: thermobarometric constraints', *New Zealand Journal of Geology and Geophysics*, 38: 525–33.

Hancox, G. T., T. J. Chinn & M. J. McSaveney, 1991. *Immediate report – Mt Cook rock avalanche, 14 December 1991*, DSIR Geology and Geophysics, Lower Hutt.

Henderson, R. D. & S. M. Thompson, 1999. 'Extreme rainfalls in the Southern Alps of New Zealand', *Journal of Hydrology (NZ)*, 38(2): 309–30.

Kamp, P. J. J., 1986. 'Late Cretaceous–Cenozoic tectonic development of the southwest Pacific region', *Tectonophysics*, 121: 225–51.

King, P. R., 2000. 'Tectonic reconstructions of New Zealand: 40 Ma to the Present', *New Zealand Journal of Geology and Geophysics*, 43: 611–38.

Kleffmann, S., F. Davey, A. Melhuish, D. Okaya & T. A. Stern, 1998. 'Crustal structure in the central South Island, New Zealand, from the Lake Pukaki seismic experiment', *New Zealand Journal of Geology and Geophysics*, 41: 39–49.

Leitner, B., D. Eberhart-Phillips, H. Anderson & J. L. Nabelek, 2001. 'A focused look at the Alpine fault, New Zealand: seismicity, focal mechanisms, and stress observations', *Journal of Geophysical Research*, 106: 2193–220.

McSaveney, M. J., T. J. Chinn & G. F. Coates, 1992. *Immediate report – the Mount Fletcher rock avalanche of 17 September 1992*, Institute of Geological & Nuclear Sciences Ltd, Lower Hutt.

Mortimer, N., 1993. *Geology of the Otago Schist and adjacent rocks. Scale 1:500,000*, Institute of Geological & Nuclear Sciences geological map 7, Institute of Geological & Nuclear Sciences Ltd, Lower Hutt.

Norris, R. J., & A. F. Cooper, 1995. 'Origin of small-scale segmentation and transpressional thrusting along the Alpine fault, New Zealand', *Geological Society of America Bulletin*, 107: 231–40.

Norris, R. J., & A. F. Cooper, 2001. 'Late Quaternary slip rates and slip partitioning on the Alpine Fault, New Zealand', *Journal of Structural Geology*, 23: 507–20.

Pierson, T. C., 1980. 'Erosion and deposition by debris flows at Mt Thomas, North Canterbury, New Zealand', *Earth Surface Processes*, 5: 227–47.

Rayners, M., 1987. 'Subcrustal earthquakes in the central South Island, New Zealand, and the root of the Southern Alps', *Geology*, 15: 1168-.

Riddolls, P. M., 1987. *New Zealand geology. Geological map of New Zealand 1:2,000,000*, DSIR, Wellington.

Smith, E. G. C., T. A. Stern & B. O'Brien, 1995. 'A seismic velocity profile across the central South Island, New Zealand, from explosion data', *New Zealand Journal of Geology and Geophysics*, 38: 565–70.

Stern, T. A., 1995. 'Gravity anomalies and crustal loading at and adjacent to the Alpine Fault, New Zealand',

New Zealand Journal of Geology and Geophysics, 38: 593–600.

Stern, T. A., P. Molnar, D. Okaya & D. Eberhart-Phillips, 2000. 'Teleseismic P-wave delays and modes of shortening the mantle lithosphere beneath South Island, New Zealand', *Journal of Geophysical Research*, 105: 21,615–31.

Stern, T., S. Kleffmann, M. Scherwath, D. Okaya & S. Bannister, 2001. 'Low seismic wave speeds and enhanced fluid pressures beneath the Southern Alps of New Zealand', *Geology*, 29: 679–82.

Suggate, R. P., 1990. 'Late Pliocene and Quaternary glaciations of New Zealand', *Quaternary Science Review*, 9: 175–97.

Sutherland, R., 1999. 'Cenozoic bending of New Zealand basement terranes and Alpine Fault displacement: a brief review', *New Zealand Journal of Geology and Geophysics*, 42: 295–301.

Sutherland, R., 1994. 'Displacement since the Pliocene along the southern section of the Alpine Fault, New Zealand', *Geology*, 22: 327–330.

Walcott, R. I., 1979. 'Plate motion and shear strain rates in the vicinity of the Southern Alps' in R. I. Walcott & H. M. Cresswell (eds), *Origin of the Southern Alps*, Royal Society of New Zealand Bulletin 18: 5–12.

Wellman, H. W., 1979. 'An uplift map for the South Island of New Zealand, and a model for the uplift of the Southern Alps', in R. I. Walcott & H. M. Cresswell (eds), *Origin of the Southern Alps*, Royal Society of New Zealand Bulletin 18: 13–20.

Wells, A., M. D. Yetton, R. P. Duncan & G. H. Stewart, 1999. 'Prehistoric dates of the most recent Alpine Fault earthquakes, New Zealand', *Geology*, 27: 995–98.

Whitehouse, I. E., 1983. 'Distribution of large rock avalanche deposits in the central Southern Alps, New Zealand', *New Zealand Journal of Geology and Geophysics*, 26: 271–79.

Woodward, D. J., 1979. 'The crustal structure of the Southern Alps, New Zealand, as determined by gravity', in R. I. Walcott & H. M. Cresswell (eds), *Origin of the Southern Alps*, Royal Society of New Zealand Bulletin 18: 95–98.

General reading

Aitken, J. J., 1996. *Plate tectonics for curious Kiwis*, Institute of Geological & Nuclear Sciences Information Series 42.

Cox, G. J., 1989. *Slumbering giants: the volcanoes and thermal regions of the central North Island*, Collins, Auckland.

Cox, G. J., & B. W. Hayward, 1999. *The restless country: volcanoes and earthquakes of New Zealand*, HarperCollins, Auckland.

Forsyth, P. J., & J. J. Aitken, 1995. *New Zealand minerals and rocks for beginners*, Institute of Geological & Nuclear Sciences Information Series 36.

Thornton, J., 1985. *Field guide to New Zealand geology: a field guide to rocks, minerals and fossils*, Reed Methuen, Auckland.

Stevens, G. R., 1980. *New Zealand adrift: the theory of continental drift in a New Zealand setting*, Reed, Wellington.

Stevens, G. R., 1985. *Lands in collision: discovering New Zealand's past geography*, SIPC, DSIR, Wellington.

GLOSSARY

Andesite   A grey fine-grained volcanic rock with medium silica content (less than rhyolite, more than basalt).

Basalt   A dark-coloured fine-grained volcanic rock.

Basement rock   Much older rock cut back by an earlier cycle of erosion, now underlying younger sedimentary rocks.

Bed   Single layer of sedimentary rock with planar boundaries separating it from layers above and below.

Biotite   One of the mica minerals, with platy form and black/brown in colour. Peels apart in thin translucent layers.

Chert   A compact rock, usually grey or creamy, consisting of microscopic crystals of silica.

Chlorite   A green mineral related to the micas.

Conglomerate   A coarse-grained sedimentary rock formed from gravel-size particles.

Continental crust   The crust of continental regions, up to 60 km thick, comprising all kinds of igneous, sedimentary and metamorphic rocks that are less dense than the oceanic crust or the underlying mantle.

Crust   The outermost layer of the Earth, 3–65 km thick.

Deformation (of rock)   The changes that occur in rock as a result of tectonic forces such as folding and faulting.

Ductile behaviour   When materials change shape without breaking.

Fault   A planar fracture in rock across which there has been displacement.

Fault zone   A tract of crushed or broken rock often found in a wide zone along a fault line.

Feldspar   Important group of rock-forming silicate minerals.

Granite   An igneous rock, coarse grained and light in colour, formed when molten rock cools at depth.

Greywacke   Hard grey sandstone.

Limestone   A sedimentary rock consisting of calcium carbonate from the hard body parts of marine organisms.

Lithosphere   The outer layer of the Earth (including the crust), about 100 km thick, which is divided into tectonic plates that move independent of one another.

Magma   Molten rock; cools to form igneous rocks.

Mantle   The bulk of the Earth from the core to the crust consisting of mainly dense iron and magnesium silicates.

Marble   A hard, usually light-coloured rock formed by the metamorphism of limestone.

Orogeny   Process that leads to the uplift of mountains through deformation of the crust.

Metamorphism   The process involving pressure and temperature that forms **metamorphic rocks** by the recrystallisation of new minerals.

Mud   Rock particle finer than 0.0625 mm.

Mudstone   A fine-grained sedimentary rock formed from mud-size particles.

Oceanic crust   That part of the Earth's crust formed at mid-ocean ridges. Consists of basalt and forms the ocean floors for a depth of 3–6 km.

Oligoclase   A type of feldspar mineral.

Olivine   A mineral, usually dark green in colour, found in igneous rocks low in silica and high in iron and magnesium.

Ophiolite   Metamorphosed igneous or sea-floor rocks, dark in colour with high iron and magnesium.

Plate tectonics   The theory and study of crustal plates; their formation, movement, interaction and destruction.

Plate   A large section of the Earth's crust and lithosphere that moves as a rigid body relative to other plates.

Pyroxene   A green mineral high in calcium and magnesium.

Radiometric dating   Method of determining the age of rock and other materials by measuring the degree of radioactive breakdown from the original parent state to daughter state.

Rhyolite   A light-coloured fine-grained volcanic rock with high silica content.

Sand   Rock particle between 0.06 and 2 mm in size

Sandstone   A medium-grained sedimentary rock formed from sand-size particles.

Schist   A metamorphic rock in which the minerals, particularly micas, have developed in fine strands

Sea-floor spreading   The process by which crustal plates move apart as magma erupts at mid-ocean ridges.

Sedimentary rock   A rock formed from particles carried and deposited by water, wind or ice

Silt   Rock particles finer than sand (< 0.06 mm) and coarser than clay (> 0.004 mm).

Subduction zone   The dipping planar zone along which one plate descends beneath another.

Trench   A long narrow trough in the ocean floor where oceanic crust dips down a subduction zone.

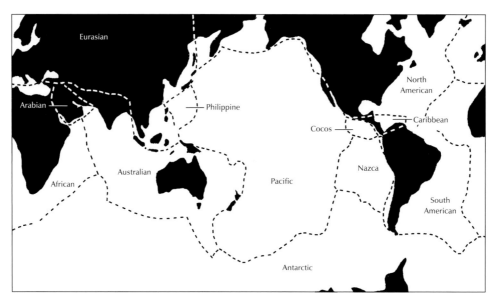

Outlines of the main tectonic plates that make up the Earth's crust.

INDEX